Applied Mathematics

Volume 1

An Introduction to Discrete Dynamical Systems and their General Solutions

4th Edition

Applied Mathematics Series Editor
Dov Gabbay   dov.gabbay@kcl.a

An Introduction to Discrete Dynamical Systems and their General Solutions

4th Edition

F. Oliveira-Pinto, Ph.D., Cantab

1st edition, 2010
2nd edition, 2011
3rd edition, 2013
4th edition, 2023

ISBN 978-1-84890-445-3

College Publications
Scientific Director: Dov Gabbay
Managing Director: Jane Spurr

http://www.collegepublications.co.uk

Cover design adapted from a figure by the author
Cover created by Laraine Welch

TABLE OF CONTENTS

Note: DS stands for *Dynamical System(s)*
          DDS for *Discrete Dynamical System(s)*
          CDS for *Continuous Dynamical System(s)*.

2-VARIABLE / POPULATION DS

INTRODUCTION to 4th EDITION

Discrete dynamical systems, in short – DDS – have been studied in the last 50 years providing very satisfying results. When we compare them to the analogue continuous ones – CDS – we find the discrete ones to be mathematically more complicated and often having solutions of a quite different nature.

Here we study 1-D and 2-D general solutions and stable equilibrium values whenever known of discrete and continuous analogue dynamical systems and identify their intrinsic differences. We start by introducing discrete dynamical systems before comparing them to continuous analogue ones. We study in some depth the chaotic-type solutions that are not present in the continuous analogue ones.

The 2nd edition contained a number of modifications and additions to the original one to make it more comprehensible and enjoyable and a new chapter on the effects of harvesting/fishing on hybrid DDS. The 3rd edition was more inter-linked on periodic solutions of quadratic DDS, on harvesting/fishing effects on oscillating quadratic DDS and on non-periodic, non-chaotic cubic DDS. Greater emphasis was placed on the stabilizing effect of harvesting/fishing on oscillating DDS. Examples of applications of DDS to accounting, pharmacology, harvesting/fishing stratrategies, population models and historical battles were given. In the 4th edition we changed the way general solutions were studied with the text edited, partly rewritten and easy to follow.

The following is based on a series of lectures given to post-graduate students at the Scuola di Dottorato (Doctoral School) Galileo Galilei at Pisa University on the same topic.

We define a 1st order Discrete Dynamical System or DDS as a sequence of real values from a function used repeatedly, often called a recurrence relation,

$$x_{k+1} = f(x_k), \qquad\qquad \text{for} \quad k = 0, 1, 2 \dots$$

1

from a given x_0. We assume that the 1st derivative of $f(x)$ in order to x, $D_x f$ exists.

A 2nd order DDS is

$$x_{k+2} = f(x_{k+1}, x_k), \qquad\qquad k = 0, 1, 2 \dots$$

from x_0 and x_1 given and, in general,

Def. We define a Discrete Dynamical System as a sequence of real values from a function used repeatedly

$$x_{k+m} = f(x_{k+m-1}, x_{k+m-2}, \dots, x_k) \qquad\qquad k = 0, 1, 2, \dots$$

from a given set of initial values $\{x_0, x_1, \dots, x_{m-1}\}$.

In the following, we often write a 1st order DDS in the form:

$$x_{k+1} = x_k + g(x_k) \qquad\qquad \text{with } x_0 \text{ given.}$$

The main advantage of this notation is that a 1st order DDS becomes a 1st order difference equation and difference equations have been studied extensively, well before DDS [we shall use DDS to mean DDS or DDS(s)] appeared on their own. In fact, we obtain

$$\Delta x_{k+1} = g(x_k) \qquad\qquad \text{with} \quad \Delta x_{k+1} = x_{k+1} - x_k,$$

which is a 1st order difference equation with an initial value x_0.

Symbols: In A → B, "→" means that B follows from A or that A leads to B; In A & B, "&" means A and B, as x_0 & y_0 in 2nd order DDS.

1. GENERAL SOLUTIONS IN CLOSED FORM

Given a 1st order DDS by its recurrence relation

$$x_{k+1} = f(x_k) \qquad \text{with } x_0 \text{ given,}$$

we can, in principle, compute the value of x_{k+1} for any integer $k > 0$ but for large k its calculation is expensive in computing time, and, more important, its numerical value may be grossly incorrect by an accumulation of arithmetic rounding errors. For instance, an accurate estimation of $\lim_{k=\infty} f(x_k)$, even if it exists, may not be possible by numerical means.

On the contrary, if we can find a non-recursive function $F(k,x_0)$ that allows us to compute x_{k+1} directly from the initial value x_0 without the knowledge of the intermediate values leading to x_k, we call $F(k,x_0)$ a general solution of the DDS and we can study it further.

As an example, let us consider the linear DDS

$$x_{k+1} = \lambda x_k \qquad \text{with } x_0 \text{ given and } \lambda > 0 \text{ constant.}$$

We shall see that

$$x_k = x_0 \lambda^k \qquad \text{or} \qquad F(k, x_0) = x_0 \lambda^k$$

is the general solution, since, for any integer $k > 0$, we can determine x_k directly from x_0.

In fact,

$$
\begin{aligned}
k = 1 &\rightarrow x_1 = \lambda x_0 \\
k = 2 &\rightarrow x_2 = \lambda \lambda x_0 = \lambda x_1 \qquad \text{and so on.}
\end{aligned}
$$

In the following we aim at the determination of general solutions of DDS for unrestricted values of the parameters, (in this case λ) to allow further study and insight. Sometimes the values of the parameters are limited to continuous intervals or only discrete real values. As we shall see, the existence of general or closed solutions is more an exception than the rule.

2. EQUILIBRIUM VALUES FOR DDS

With dynamical systems, the knowledge of their equilibrium values is essential since, on them, they may come to a steady state.

Let a 1st order DDS be given by,

$$x_{k+1} = f(x_k),$$
with x_0 given.

Def.   A value x^* is said to be an equilibrium value if, at x^*, $f(x^*)$ reproduces x^* exactly, i.e. $x^* = f(x^*)$.

Next, we have to find if these equilibrium values x^* are stable or unstable, i.e. whether or not a small perturbation Δx near x^* produces a sequence of values $\{x_k\}$ that tends to x^* or not, respectively.

To study the stability of DDS at the equilibrium values, let us consider a Taylor series of $f(x)$ around x^* with a 1st order Lagrange reminder:

$$f(x) = f(x^*) + (D_x f)_{x=\eta}(x - x^*), \qquad \eta \subset (x^*, x).$$

If $D_x f$ exists and $|D_x f| < 1$ for x in the neighbourhood of x^* (both right and left neighbourhood of x^*),

$$|f(x_k) - f(x^*)| < |x_k - x^*|$$
and
$$|x_{k+1} - x^*| < |x_k - x^*|$$

for all $k > 0$ and the equilibrium value is called stable.

For values of $|D_x f| > 1$ around x^*,

$$|x_{k+1} - x^*| > |x_k - x^*|$$

for increasing k and the equilibrium value is said to be unstable.

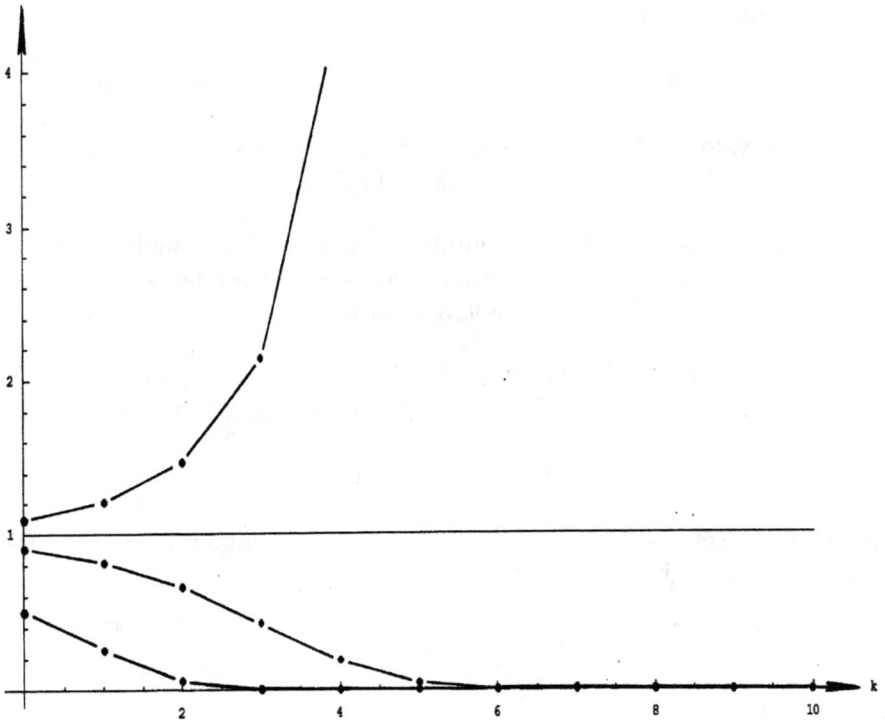

Fig. 1 Stable and unstable solutions

6

The equilibrium value x* is said to be unstable since minute perturbations in the value of x near x* lead the DDS to move away from it.

The special case of $| D_x f | = 1$ is inconclusive. In theory,

$$| x_{k+1} - x^* | = | x_k - x^* |$$

and the solution does not approach or move away from x* for increasing k .

However, in practice, we have to assume $| D_x f | = 1$ not only at x* but also **constant** and **equal** to **1** in an interval around x* and this is not usually the case. In fact, the DDS may be stable to the right of x* and not to the left etc. and we do not consider this special case on its own.

Thus, we need to find the equilibrium values first by solving the equation

$$x - f(x) = 0 .$$

When f (x) is a polynomial in x, the job is simplified since the x* are the zeros of the polynomial $x - f(x)$ and, at least, their number is known. Otherwise, real equilibrium values may not even exist. Note that only the real zeros are relevant to us.

Ex. 1:   Let us consider the DDS

$$x_{k+1} = [x_k]^2 \qquad\qquad \text{with } x_0 \text{ given.}$$

Equilibrium values are the zeros of

$$x - f(x) = x[x - 1] \qquad \text{with} \qquad f(x) = x^2$$

and the equilibrium values are: $x_1* = 0$ and $x_2* = 1$.

Regarding stability, the derivatives in x, $D_x f$ are,

$$(D_x f)_{x=0} = 0 \qquad \text{and} \qquad (D_x f)_{x=1} = 2$$

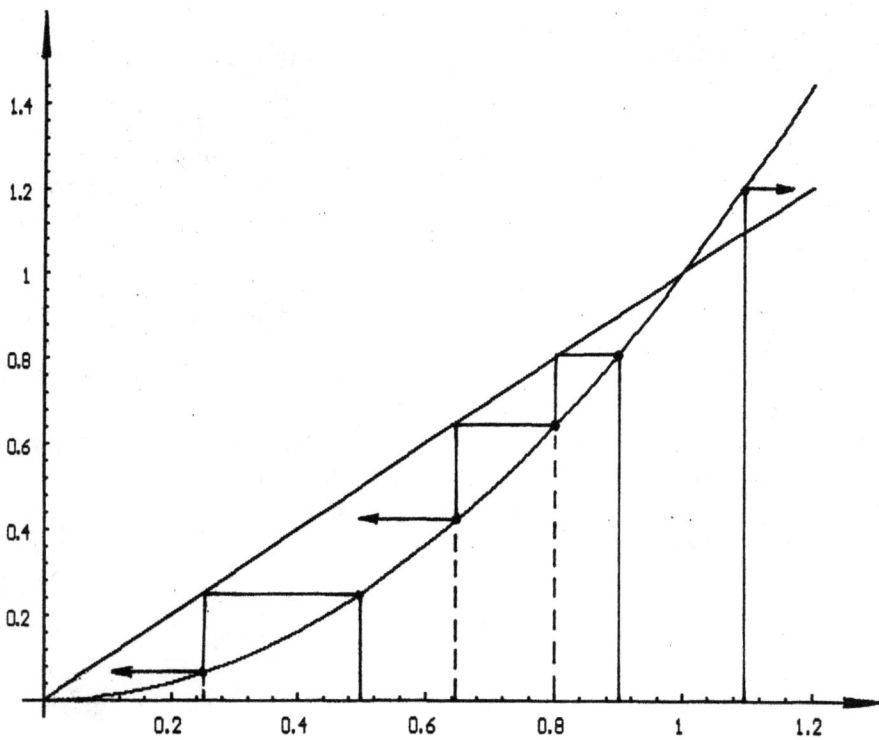

Fig. 2 Alternative representation of stable and unstable solutions

and the equilibrium values are of different nature: x_1^* stable and x_2^* unstable. Fig.1 has three discrete solutions with different $x_0 = 0.5$, $x_0 = 0.9$ and $x_0 = 1.1$ or $x_0 = 0.5, 0.9, 1.1$. Note that all solutions move away from $x_2^* = 1$. Two approach $x_1^* = 0$.

Historically, long before digital computation was widely available, the study of equilibrium values was done graphically by considering the system of relations:

$$\begin{cases} y_{k+1} = f(x_k) \\ \\ x_{k+1} = y_{k+1} \; . \end{cases}$$

The equilibrium values are the intersections of the curve $y = f(x)$ with the straight line $y = x$. Fig. 2 has the same solutions with $x_0 = 0.5, 0.9, 1.1$ in a different way.

Let us now consider a DDS given by:

$$x_{k+1} = f(f(x_k)) \qquad\qquad \text{with } x_0 \text{ given}$$

which for k finite has only half of the values $\{x_k\}$ of the original

$$x_{k+1} = f(x_k) \; .$$

The equilibrium values x^* that make $x^* = f(x^*)$ are still equilibrium values of $x_{k+1} = f(f(x_k))$ since $x^* = f(f(x^*))$ but other equilibrium values may exist and of a different nature.

In fact, if $f(x)$ is a polynomial of degree n in x, $f(f(x)) - x$ is a polynomial of degree $2n$ in x. So, if some of the new zeros are real,they are equilibrium values too (stable or unstable) and are called equilibrium values of period 2 of $x_{k+1} = f(x_k)$ as *Ex.* 2 bellow makes it clearer.

So, we may have two types of equilibrium values: the originals $x^* = f(x^*)$ and the period 2 ones $x^{**} = f(x^*)$ and $x^* = f(x^{**})$ that combined give $x^* = f(f(x^*))$ or $x^{**} = f(f(x^{**}))$ exactly.

9

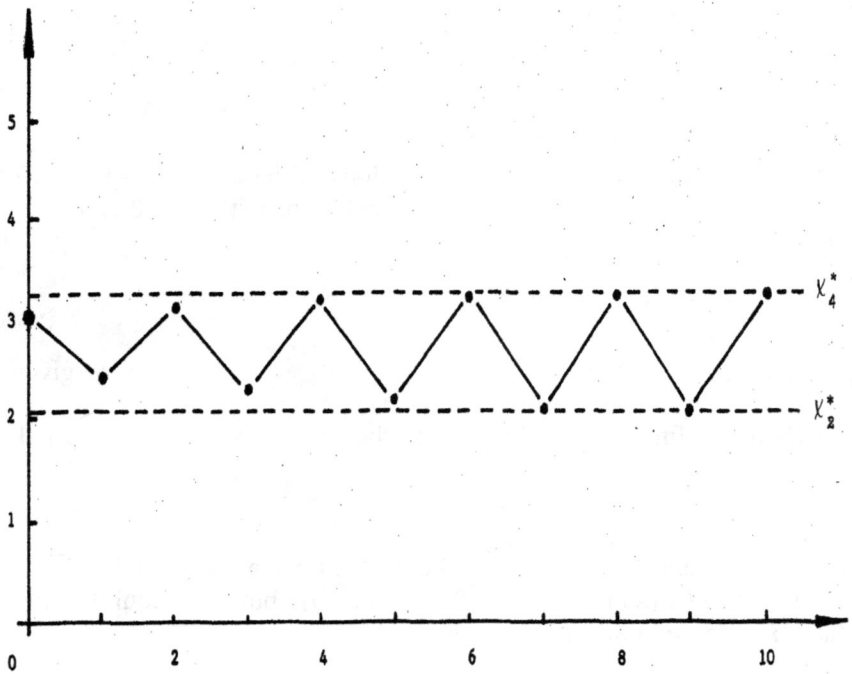

Fig. 3 Periodic solution of period 2

10

Regarding stability, we still need:

$$| D_x f(f) | < 1 \quad \text{or} \quad | D_f f | \, | D_x f | < 1$$

now with the individual derivatives defined in the neighbourhood of x^{**} and x^* respectively.

Note that when $x^* = x^{**}$, the two individual derivatives if not equal to 1 are either both < 1 or both > 1 and

$$| D_f f | \, | D_x f | < 1 \quad \text{or} \quad | D_f f | \, | D_x f | > 1 .$$

So, if x^* is stable for the original DDS $x_{k+1} = f(x_k)$ then it is still stable for $x_{k+1} = f(f(x_k))$ and, if unstable for the former it is still unstable for the latter.

The situation is more interesting when $x^* \neq x^{**}$.

Ex. 2: Let us now consider the DDS to be defined by:

$$x_{k+1} = 16/5 \, x_k - 4/5 \, [x_k]^2 \qquad \text{with} \quad x_0 \text{ given.}$$

From $\quad x = f(x) \quad$ with $\quad f(x) = 4/5 \, [4x - x^2] \quad$ we have

$$x - f(x) = -x \, [11 - 4x] / 5$$

and the equilibrium values are: $\quad x_1^* = 0 \quad$ and $\quad x_2^* = 2.75$.

Since $\quad | (D_x f)_{x=0} | = 3.2 \quad$ and $\quad | (D_x f)_{x=2.75} | = 1.2$

both equilibrium values are unstable.
However, if we consider the DDS to be written as:

$$x_{k+2} = f(f(x_k)) \qquad \text{with} \qquad f(x) = 4/5 \, [4x - x^2]$$

or

$$x - f(f(x)) = -x \, [11 - 4x] \, [105 - 84x + 16x^2] / 125$$

we have, as equilibrium values,

$$x_1* = 0, \quad x_2* = 2.052..., \quad x_3* = 2.75, \quad x_4* = 3.198...$$

with x_1* and x_3* still unstable but x_2* and x_4* stable since

$$|(D_x f)_{x=2.052}| \; |(D_f f|)_{f=f(2.052)}| = 0.159 < 1$$

or

$$|(D_x f)_{x=3.198}| \; |(D_f f|)_{f=f(3.198)}| = 0.159 < 1 .$$

So, we have the two previous x_1* and x_3* unstable plus two new periodic stable ones x_2* and x_4*.

From Fig. 3 with $x_0 = 3$, we guess that the quadratic DDS converges to a stable 2-value solution: $x_2* = 2.052...$ and $x_4* = 3.198...$ which is called a periodic solution of period 2 or a 2-point equilibrium value and it is stable.

Fig. 4 shows it more vividly when we use the system of relations

$$\begin{cases} y_{k+1} = 16/5 \; x_k - 4/5 \; [x_k]^2 \\ \\ x_{k+1} = y_{k+1} \end{cases} \qquad \text{with} \quad x_0 = 3.$$

Similarly, for periodic solutions of period 3, we would consider the equilibrium values of

$$x_{k+1} = f(f(f(x_k)))$$

that are not the equilibrium values of the original $x_{k+1} = f(x_k)$, and so on.

12

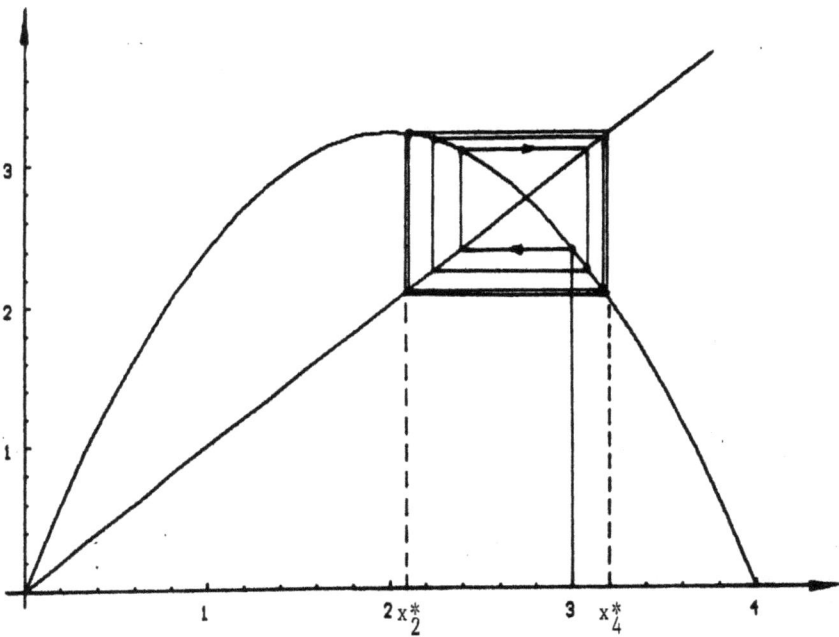

Fig. 4 Alternative representation of a periodic solution

3. SIMPLE DDS WITH EXPONENTIAL GROWTH

The simplest 1st order DDS show exponential growth. We shall consider 4 cases.

Case 1: Let us start with a linear DDS of the form, $\lambda > 0$ constant, x_0 given

$$x_{k+1} = \lambda x_k .$$

By repeated substitution, the general solution is:

$$x_k = \lambda x_{k-1} = \lambda^2 x_{k-2} = \ldots = \lambda^k x_0 \qquad \text{for all } \lambda > 0 .$$

Note that for $\lambda < 1$ the solution tends to the equilibrium $x^* = 0$ and it is stable. For $\lambda > 1$ no stable equilibrium value exists.

Case 2: More generally, consider a linear DDS with $\lambda, \mu > 0$ constant, x_0 given

$$x_{k+1} = \lambda x_k + \mu .$$

By substitution,

$$x_1 = \lambda x_0 + \mu$$
$$x_2 = \lambda^2 x_0 + \lambda \mu + \mu$$
$$x_3 = \lambda^3 x_0 + \lambda^2 \mu + \lambda \mu + \mu$$
$$\cdots$$
$$x_k = \lambda^k x_0 + \mu [1 + \lambda + \lambda^2 + \ldots + \lambda^{k-1}]$$

or summing the geometric progression

$$x_k = \lambda^k x_0 + \mu [1 - \lambda^k] / [1 - \lambda] .$$

So, the general solution is still exponential and given by

$$x_k = [x_0 - c] \lambda^k + c \qquad \text{with} \qquad c = \mu / [1 - \lambda] , \qquad \lambda \neq 1 .$$

An alternative way to calculate this general solution (which we shall use in the future) is to try, as possible general solution, a guessed expression of the form

$$x_k = b\lambda^k + c$$

where the unknown constants b and c are to be determined later.

Replacing it into the left and right hand side of the recurrence relation above, we obtain

$$b\lambda^{k+1} + c = b\lambda^{k+1} + c\lambda + \mu .$$

Simplifying

$$c = c\lambda + \mu \qquad \rightarrow \qquad c = \mu / [1-\lambda] .$$

From the guessed solution, for $k = 0$,

$$x_0 = b + c \qquad \rightarrow \qquad b = x_0 - c .$$

So, replacing both constants, we obtain

$$\mathbf{x_k = [x_0 - c]\ \lambda^k + c} \qquad \text{with} \qquad \mathbf{c = \mu / [1 - \lambda]} ,$$

which is the previous general solution $\lambda \neq 1$.

For $\lambda < 1$ the solution tends to a non-zero equilibrium $x^* = c$. For $\lambda > 1$ no stable equilibrium exists. In the special case of $\lambda = 1$, the general solution can be found to be:

$$\mathbf{x_k = x_0 + \mu\, k} \qquad\qquad \text{with } x_0 \text{ given}$$

and it is linear instead of exponential.

Case 3: Let us consider now a non-linear DDS with $\lambda, \mu, r > 0$ and $r \neq \lambda$,

$$x_{k+1} = \lambda x_k + \mu r^k \qquad\qquad \text{with } x_0 \text{ given.}$$

We are going to look for the general solution by assuming, from *Case* 2, to be of the form

$$x_k = b\lambda^k + cr^k$$

where the constants b and c are to be determined later.

Replacing it into the left and right hand side of the recurrence relation,

$$b\lambda^{k+1} + cr^{k+1} = b\lambda^{k+1} + c\lambda r^k + \mu r^k.$$

Simplifying it,

$$cr = c\lambda + \mu \qquad \rightarrow \qquad c = \mu/[r - \lambda].$$

Also, from our guessed solution for k = 0,

$$x_0 = b + c \qquad \rightarrow \qquad b = x_0 - c$$

and our job is done.

So, the general solution with x_0 given is exponential and written as:

$$\mathbf{x_k = [x_0 - c]\,\lambda^k + cr^k,} \qquad \lambda, r > 0.$$

Note that this DDS has no stable equilibrium for $\lambda > 1$ or $r > 1$.

Case 4: Now let us study the previous DDS when $r = \lambda$, and $\lambda, \mu > 0$,

$$x_{k+1} = rx_k + \mu r^k \qquad\qquad \text{with} \quad x_0 \text{ given.}$$

The guessed solution is, in this case, taken as:

$$x_k = [b + ck]\,r^k$$

with the constants b and c to be determined later.

Replacing it into the recurrence relation

$$[\,b + c\,[\,k+1\,]\,]\ r^{k+1} = [b + c\,k]\ r^{k+1} + \mu\,r^{k}.$$

We obtain, by simplification,

$$c\,r = \mu \qquad \rightarrow \qquad c = \mu/r.$$

Also for $k = 0$, we have,

$$x_0 = b \qquad \rightarrow \qquad b = x_0.$$

So, with x_0 given, the general solution is still exponential with

$$x_k = x_0\ r^{k} + \mu\,k\ r^{k-1}.$$

Again, this DDS has no stable equilibrium for $r > 1$.

4. BANK COMPOUND INTEREST AS A DDS

On a deposit bank account, interest on the deposit is paid after a certain period of time, say 1 year.

When that happens we say that a "conversion" takes place and the original capital or "principal" is increased by the interest earned during that period. For the next period, the interest will be earned on the new "principal" and so on.

So, in the case when several "conversions" take place yearly, we denote the fractional interest in between "conversions" by I/T where I is the annual interest (which we assume to be constant) and $1/T$ the time interval between "conversions". Thus, $T = 12$ for monthly "conversions", $T = 365$ for daily "conversions" and so on.

Then, for a constant annual interest and no capital withdrawn, the DDS is defined by the linear recurrence relation

$$x_{k+1} = [1 + I/T] \, x_k \qquad \text{with} \quad I, T, x_0 > 0 \text{ given}$$

where x_k is the capital on the account after k "conversions" and x_{k+1} the capital at the following "conversion".

From *Case 1* of the previous chapter ($\lambda = 1 + I/T$), the general solution of this linear recurrence relation is:

$$\mathbf{x_k = x_0 \, [1 + I/T]^k}$$

which is a process of exponential growth.

Note that after one year T "conversions" have taken place and the capital on the account is

$$x_T = x_0 \, [1 + I/T]^T \qquad \text{with} \quad x_0 \text{ the starting capital.}$$

This expression is known in banking circles as the *compound interest* formula.

As an application, let us find the fixed annual interest rate that guarantees to duplicate the starting capital without withdrawals in a given time interval, say 10 years.

We have for T = 1

$$2\,x_0 = x_0\,[1+I]^{10}$$

or

$$I = e^{\ln(2)/10} - 1 \qquad \rightarrow \qquad I = 0.07177...$$

i.e. a mere 7.18 % annual interest rate will do it.

In a similar way, it can be shown that a 14.87 % annual interest rate will quadruplicate the starting capital in 10 years time! And so on.

5. MORTGAGE REPAYMENT AS A DDS

In a mortgage account, the repayments R (here assumed to be constant) take place at regular intervals while the interest I (also assumed to be constant) is added, also periodically, to the outstanding loan.

If "conversions" take place at the same time that repayments are made (which, incidentally, is not usually the case) this DDS can be described by the linear recurrence relation:

$$x_{k+1} = [1 + I/T] \ x_k - R \qquad \text{with} \quad I, T, R, x_0 \ \text{given,}$$

where $x_k > 0$ is the capital of the loan outstanding at the previous "conversion" and repayment, I the annual interest, I/T the fractional interest in between repayments and R the periodic mortgage repayment. Further, I, T, R are all assumed to be constant and positive during the duration of the loan.

So, as soon as x_{k+1} becomes zero (or negative) the DDS comes to a halt. From *Case 2* in the chapter prior to previous one, the general solution of this linear recurrence relation with $\mu = -R$, $\lambda = 1 + I/T$ and $c = RT/I$ is

$$\mathbf{x_k = [\ x_0 - R\ T/I\ ] \ [1 + I/T]^k + R\ T/I}$$

which is known as the *mortgage repayment* formula.

How comes that an exponential solution (in fact $\lambda > 1$) describes an essentially decreasing process ? In effect, if we choose the repayment R such that

$$R > x_0\ I/T \qquad \rightarrow \qquad [\ x_0 - RT/I\ ] < 0$$

and the x_k decreases to zero.

Frequently, we have to find R such, that for a given k, say $k = N$ we have $x_N = 0$ (or x_N nearly zero). To that effect, take *mortgage repayment* formula above and solve it for R with $x_N = 0$.

We get

$$R = q\, x_0 \,/\, [\, 1 - [1 + q]^{-N} \,] \qquad\qquad \text{with} \qquad\qquad q = I/T .$$

As an example, let us find the constant repayment R for a 5 year loan of $x_0 = £/€\,10\,000$ at a fixed interest of 6 % paid monthly. We have

$$N = 60 , \; I = 0.06, \; T = 12 \qquad \rightarrow \qquad R = £/€\,193.33 .$$

6. CONTINUOUS / DISCRETE DS

Now, the aim is to compare CDS - Continuous Dynamical System(s) with DDS ones.

We recall the bank compound interest DDS as

$$x_{k+1} = \lambda x_k \qquad\qquad \text{and} \qquad \lambda = 1 + I/T$$

with x_0 given and $I, T > 0$ constant.

It can also be expressed as a difference equation:

$$\Delta x_{k+1} = [\lambda - 1] x_k \qquad\qquad \text{with} \qquad \Delta x_{k+1} = x_{k+1} - x_k .$$

The corresponding CDS is defined by the differential equation:

$$D_t y = [\lambda - 1] y \qquad\qquad \text{with } y_0 = y(t_0) \text{ given.}$$

Here, we use the variable y instead of the variable x to help distinguishing between CDS(s) and DDS(s) respectively.

The general solution of this linear CDS $y_0 = y(t_0)$ is (see ahead):

$$\mathbf{y} = \mathbf{y_0} \ \mathbf{e}^{[\lambda-1][t-t_0]} \qquad\qquad \text{for } t \geq t_0 ,$$

which is known as the *Malthusian Law of Growth* and corresponds to the bank *compound interest* when the interest is added continuously to the capital.

In general, for a DDS defined by

$$x_{k+1} = x_k + f(x_k) \qquad\qquad \text{with } x_0 \text{ given,}$$

the associated CDS is a real time function y(t), $t \geq t_0$ solution of the differential equation [f(x) the same as f(y) only y replaces x]

$$D_t y = f(y) \qquad\qquad \text{with } y_0 = y(t_0) \text{ given.}$$

Def. A value y* is said to be an equilibrium value at infinity, or simply an equilibrium value of y(t) if $\lim_{t=\infty} y = y*$ exists.

7. MALTHUSIAN LAW OF GROWTH

To obtain the *Malthusian law of growth* we integrate, $\lambda > 0$ constant,

$$D_t y = [\lambda - 1] y \qquad\qquad \text{with} \quad y_0 = y(t_0) \text{ given .}$$

We have

$$D_t y / y = \lambda - 1 .$$

Integrating

$$\ln y = [\lambda - 1] t + K^{\#}$$

or

$$y = K e^{[\lambda - 1] t}$$

with

$$K = y_0 e^{-[\lambda - 1] t_0} .$$

Replacing

$$\mathbf{y = y_0 \ e^{[\lambda - 1][t - t_0]}}$$

which is a exponential type solution, y_0 given and called the *Malthusian law of growth*.

8. BOUNDED LINEAR DDS

Now we shall introduce a limitation on the growth mechanism of previous linear DDS such that the long-term solutions may be bounded and non-zero. We shall also refer them as saturation linear DDS.

A linear alternative to the previous ones is a DDS of the form

$$x_{k+1} = x_k + \varepsilon [x_\infty - x_k] \qquad \text{with } x_0 \text{ given}$$

and $\varepsilon, x_\infty > 0$ constant. Note that the "growth rate" decreases for increasing x_k (assuming that $x_0 < x_\infty$), vanishing for $x_k = x_\infty$ (provided there is a x_k with the exact value of x_∞).

This recurrence relation can also be written as

$$x_{k+1} = [1 - \varepsilon] x_k + \varepsilon x_\infty$$

which is a linear DDS of a type previously studied.

The general solution is the exponential one for all $\varepsilon > 0$

$$x_k = [x_0 - c] [1 - \varepsilon]^k + c \qquad \text{with} \quad c = x_\infty , x_\infty \neq x_0 .$$

Replacing and re-arranging we have,

$$x_k = x_\infty - [x_\infty - x_0] [1 - \varepsilon]^k .$$

So, for $|1 - \varepsilon| < 1$ or $0 < \varepsilon < 2$ the solution tends to the equilibrium value $x^* = x_\infty$. Otherwise, for $\varepsilon > 2$ no stable equilibrium exists.

Fig(s) 5, 6, 7 show saturation linear DDS with the same $x_0 = 0.25$ and $x_\infty = 1$ but different $\varepsilon = 0.5$, 1.5, 2.5 . Note that the one with $\varepsilon = 2.5$ displays exponential growth. No longer of bounded behaviour.

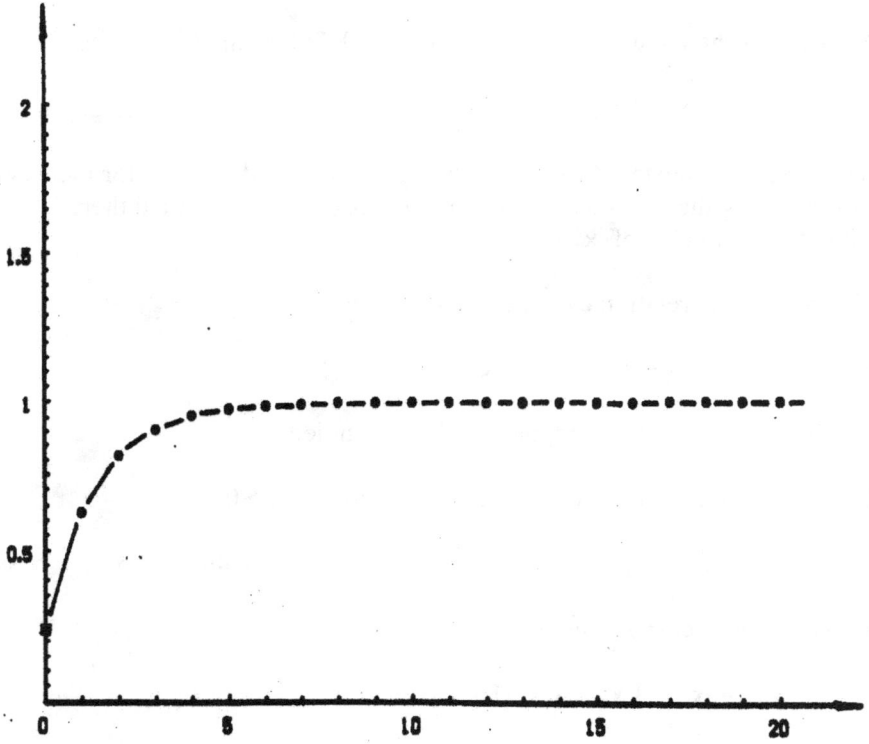

Fig. 5 Saturation linear DDS with $\varepsilon = 0.5$

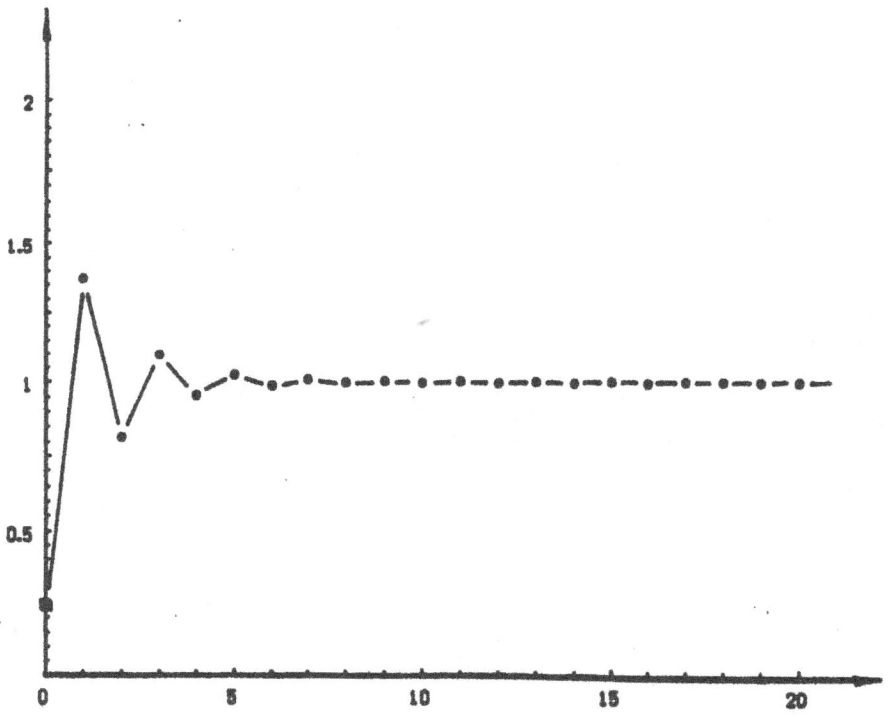

Fig. 6 Saturation linear DDS with ε = 1.5

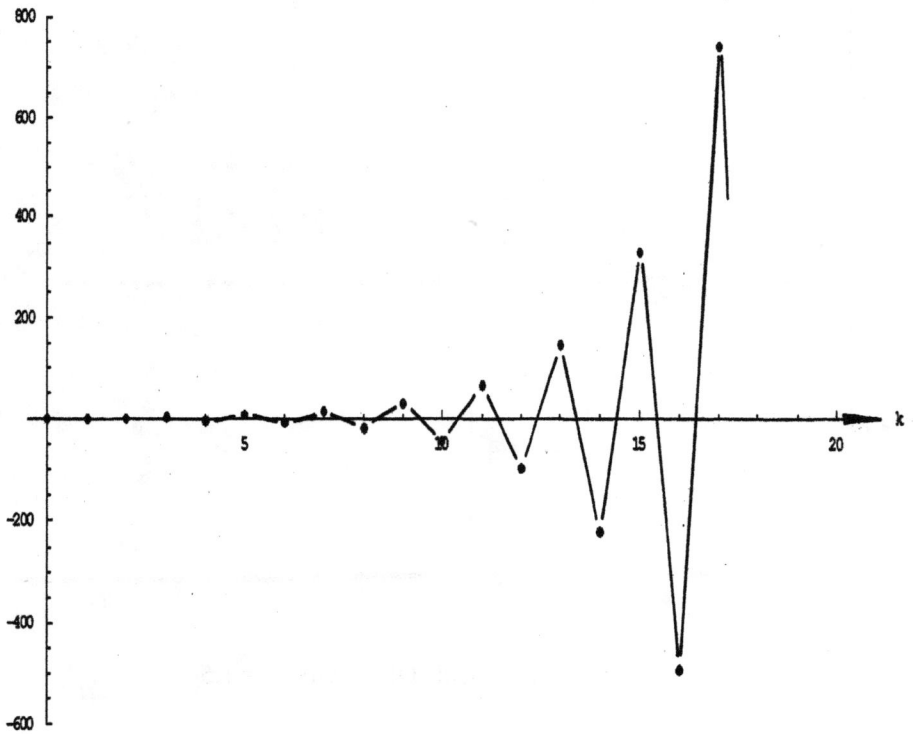

Fig. 7 Saturation linear DDS with ε = 2.5

9. BOUNDED LINEAR CDS

The continuous analogue of the previous DDS

$$x_{k+1} = x_k + \varepsilon [x_\infty - x_k]$$

or

$$\Delta x_{k+1} = \varepsilon [x_\infty - x_k]$$

is

$$D_t y = \varepsilon [y_\infty - y]$$

with $y_0 = y(t_0)$ given and ε, $y_\infty > 0$ constant.

The general solution of this linear differential equation is,

$$y = y_\infty - [y_\infty - y_0] e^{-\varepsilon [t - t_0]}$$

which has a very smooth geometric behaviour for increasing values of $\varepsilon > 0$ with a stable equilibrium value $y^* = y_\infty$.

If we compare this general solution with the equivalent DDS one

$$x_k = x_\infty - [x_\infty - x_0] [1 - \varepsilon]^k ,$$

we find that both are exponential solutions with the same analytical structure only the base of the exponentials is different. In the discrete case the solution is unbounded for $\varepsilon > 2$.

In Fig. 8 we have 3 linear CDS with the same $y_0 = 0.25$ and $y_\infty = 1$ but different $\varepsilon = 1, 2, 3$. The solutions tend always to the stable equilibrium value $y^* = y_\infty$ which is independent of ε and y_0 .We may also say that the solutions grow asymptotically to a saturation value y_∞ .

So, for values of $\varepsilon > 2$ the behaviour of these CDS is very different from the discrete counterparts previously studied.

31

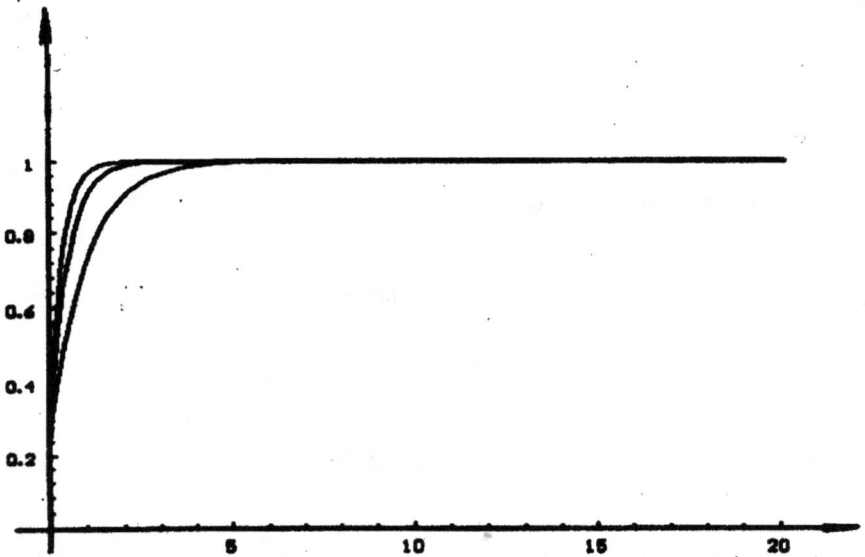

Fig. 8 Saturation linear CDS with ε = 1, 2, 3

32

10. SOLUTION OF BOUNDED CDS

In the following, with ε, $y_\infty > 0$ constant, we find the general solution of the differential equation,

$$D_t\, y \;=\; \varepsilon\,[y_\infty - y] \qquad\qquad \text{with } y_0 = y(t_0) \text{ given .}$$

Separating variables,

$$[y_\infty - y]^{-1}\, dy \;=\; \varepsilon\, dt\, .$$

Integrating,

$$\ln(y_\infty - y) \;=\; -\,\varepsilon\, t\; +\; K^{\#}$$

or

$$y \;=\; y_\infty - K\, e^{-\varepsilon t}$$

with

$$K = [y_\infty - y_0]\; e^{\varepsilon t_0}\, .$$

Replacing, we have with y_0 given, the general solution,

$$\mathbf{y} \;=\; \mathbf{y_\infty}\; -\; [\mathbf{y_\infty} - \mathbf{y_0}]\, \mathbf{e}^{-\varepsilon\,[\,t - t_0\,]}$$

as previously stated.

11. PHARMACOLOGICAL EXAMPLE

Now we consider a pharmacology case study which is not strictly a continuous or discrete dynamical system but a piecewise continuous one. It behaves like a bounded linear DDS.

This example is about the quantity of medication required to achieve a certain drug concentration level in a patient's blood when the medication is taken repeatedly at constant time intervals T $(T > 0)$.

First, we assume that the concentration of the drug in the patient's blood stream as a function of time t can be described by a simple linear differential equation

$$D_t y = -\varepsilon y \qquad\qquad \text{with } y_0 = y(t_0) \text{ and } \varepsilon > 0.$$

The coefficient $\varepsilon > 0$ is assumed time constant and only dependent on the type of drug used.

Being a linear differential equation, its general solution is

$$y = y_0 e^{-\varepsilon[t-t_0]}$$

with y_0 the medication taken at time t_0 and, in the long-term, $\lim_{t=\infty} y = 0$.

Consequently, assume that at time $t_1 = t_0 + T$, the patient takes another dose y_0 identical to the one at time t_0. So, at time t_1, in the blood stream there is old and fresh drug amounts and the new differential equation initial condition is
$$y_1 = y_0[1 + e^{-\varepsilon T}] > y_0.$$

So, for increasing $t \geq t_1$, the drug concentration in the patient blood stream is

$$y = y_0[1 + e^{-\varepsilon T}] e^{-\varepsilon[t-t_1]}.$$

Similarly, at time $t_2 = t_0 + 2T$, another dose of y_0 is given to the patient and the differential equation initial condition becomes

$$y_2 = y_0 \, [1 + e^{-\varepsilon T} + e^{-2\varepsilon T}] \; > y_1 \; .$$

Then, after k+1 dosages y_0 at constant time intervals T, the differential equation initial condition is

$$y_k = y_0 \, [1 + \Sigma_i \, e^{-i\varepsilon T}] \; , \qquad\qquad i = 1,2,\ldots k \; .$$

The geometric progression has the sum,

$$1 + \Sigma_i \, e^{-i\varepsilon T} = [1 - e^{-[k+1]\varepsilon T}] \, / \, [1 - e^{-\varepsilon T}]$$

and

$$y_\infty = \lim_{k=\infty} y_k = y_0 \, / \, [1 - e^{-\varepsilon T}]$$

y_∞ being called the saturation dosage peak level.

Fig. 9 shows a schematic solution of the differential equation in this case study. From a medical point of view, the inverse problem may be also relevant, i.e. given y_∞ find a suitable y_0.

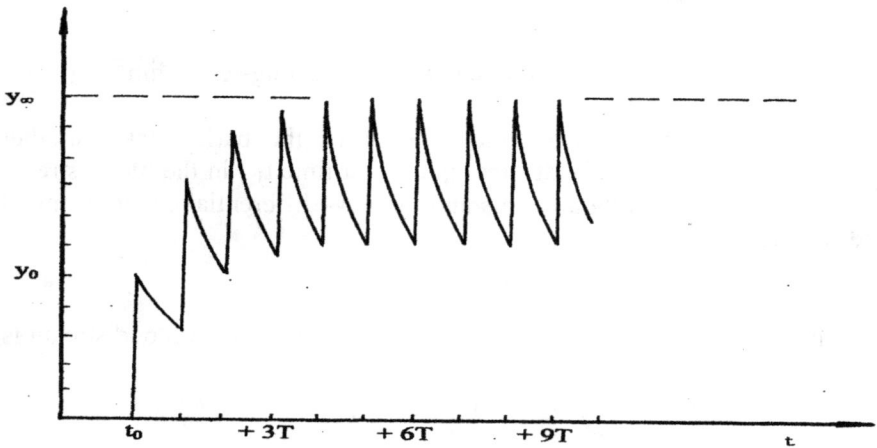

Fig. 9 Saturation example in pharmacology

12. QUADRATIC DDS

As the next growth model we would like to consider a quadratic DDS written as

$$x_{k+1} = x_k + \varepsilon [x_\infty - x_k] \, x_k$$

with a "saturation" factor $[x_\infty - x_k]$ **and** a "growth" one εx_k, $\varepsilon, x_\infty > 0$.

However, to have a more flexible notation for future reference, we shall take it as

$$x_{k+1} = x_k + \varepsilon [1 - \gamma x_k] \, x_k$$

or

$$\Delta x_{k+1} = \varepsilon [1 - \gamma x_k] \, x_k \qquad \text{with } x_0 \text{ given}$$

and with the "growth" and the "attrition" coefficients $\varepsilon, \gamma > 0$ constant. Note that the "attrition" coefficient γ is equivalent to $1/x_\infty$.

The equilibrium values are $x^* = 0$ and $x^* = \gamma^{-1}$ and the "growth rate" for x_k should decrease to zero (assuming that $0 < x_0 < x^*$) as x_k tends to $x^* = \gamma^{-1}$. Remember, however, that x_k being a discrete variable may never get one value to the exact value of γ^{-1}, i.e. the "growth rate" may not decrease to zero as we would expect it to do.

With

$$f(x) = x + \varepsilon [1 - \gamma x] \, x \ ,$$

$$(D_x f)_{x^*=0} = 1 + \varepsilon > 1 \quad \text{and} \quad (D_x f)_{x^*=1/\gamma} = 1 - \varepsilon$$

and the equilibrium values are always unstable at the origin but stable at $x^* = \gamma^{-1}$ for $|1 - \varepsilon| < 1$ or $0 < \varepsilon < 2$.

Again, regarding stability, the value of the "growth" coefficient ε is critical.

In fact, Fig(s) 10, 11, 12, 13 show quadratic DDS with the same $x_0 = 0.25$,

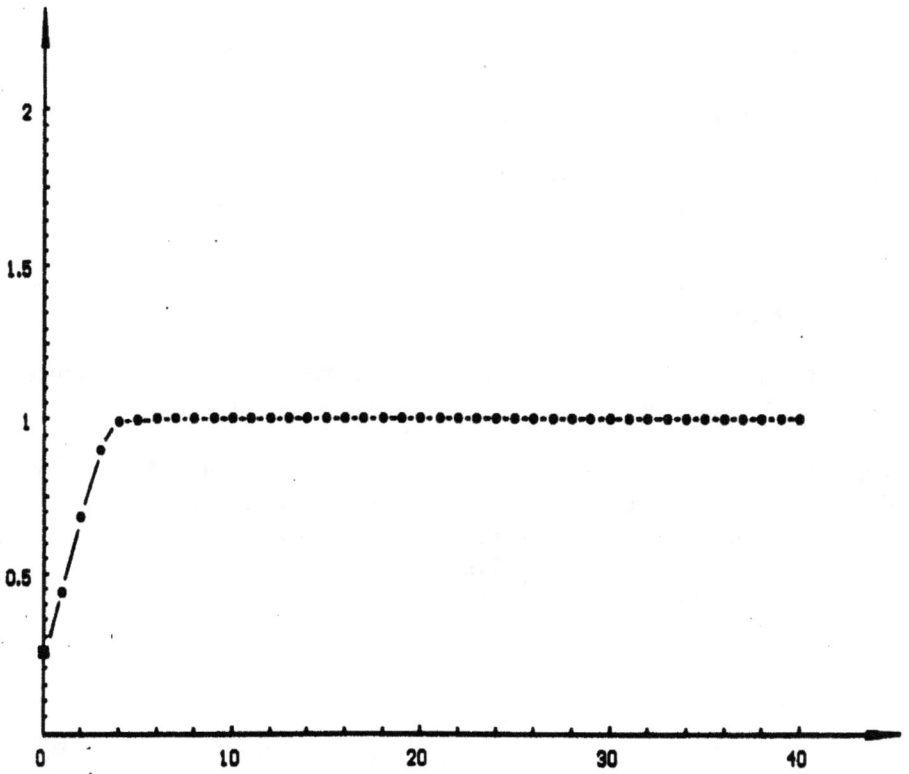

Fig.10 Quadratic DDS with $\varepsilon = 1, \gamma = 1$

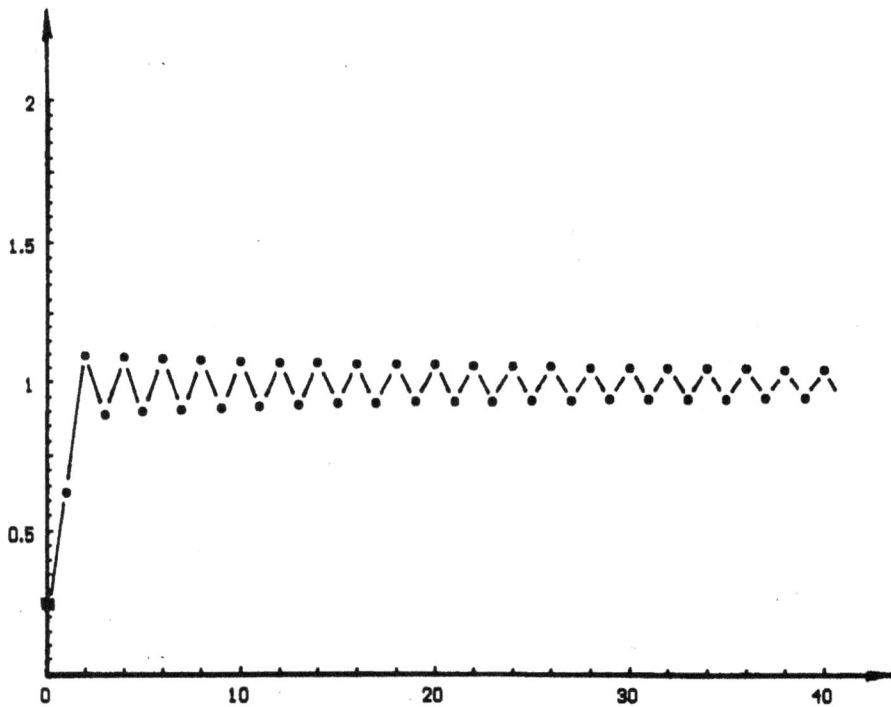

Fig. 11 Quadratic DDS with $\varepsilon = 2, \gamma = 1$

$\gamma = 1$ but different values for ε. The solutions change dramatically from "smooth" $\varepsilon = 1$ in Fig. 10, not so "smooth" $\varepsilon = 2$ in Fig. 11, to periodic $\varepsilon = 2.5$ in Fig. 12 and finally "chaotic" but bounded $\varepsilon = 3$ in Fig.13. Periodic solutions are typical of non-linear DDS.

Regarding periodic solutions, let us look into 2-point periodic solutions of

$$x_{k+1} = f(x_k) \qquad \text{with} \qquad f(x_k) = x_k + \varepsilon [1 - \gamma x_k] x_k .$$

They are single point solutions of

$$x_{k+2} = f(f(x_k))$$

i.e.

$$x - f(f(x)) = -\varepsilon [\varepsilon + 2] x + \varepsilon \gamma [\varepsilon + 1] [\varepsilon + 2] x^2 - 2 \varepsilon^2 \gamma^2 [\varepsilon + 1] x^3 + \varepsilon^3 \gamma^3 x^4 .$$

This function has the zeros $x_1^* = 0$ and $x_2^* = \gamma^{-1}$ of the original $f(x_k)$

$$x_{1,2}^* = f(f(x_{1,2}^*))$$

plus two new ones x_3^* and x_4^* since it is a polynomial of degree 4 in x. So it can be written as

$$x - f(f(x)) = x [x - x_2^*] [a x^2 + b x + c] .$$

For the new x_3^* and x_4^* to be real and distinct we must have $b^2 - 4ac > 0$. Equating coefficients of likely powers and making $x_2^* = \gamma^{-1}$, we have

$$a = \varepsilon^3 \gamma^3$$
$$b = -\varepsilon^2 \gamma^2 [\varepsilon + 2]$$
$$c = \varepsilon \gamma [\varepsilon + 2]$$

and the two zeros of $a x^2 + b x + c$ with a, b, c given above are

$$x_{3,4}^* = \tfrac{1}{2} \varepsilon^{-1} \gamma^{-1} [\varepsilon + 2]^{1/2} [[\varepsilon + 2]^{1/2} \pm [\varepsilon - 2]^{1/2}] .$$

These zeros are the the 2-point periodic solution of the original $f(x_k)$ and, more important, to be real and distint we must have $\varepsilon - 2 > 0$ or $\varepsilon > 2$.

Regarding the stability of the 2-point periodic solution, we must find if

40

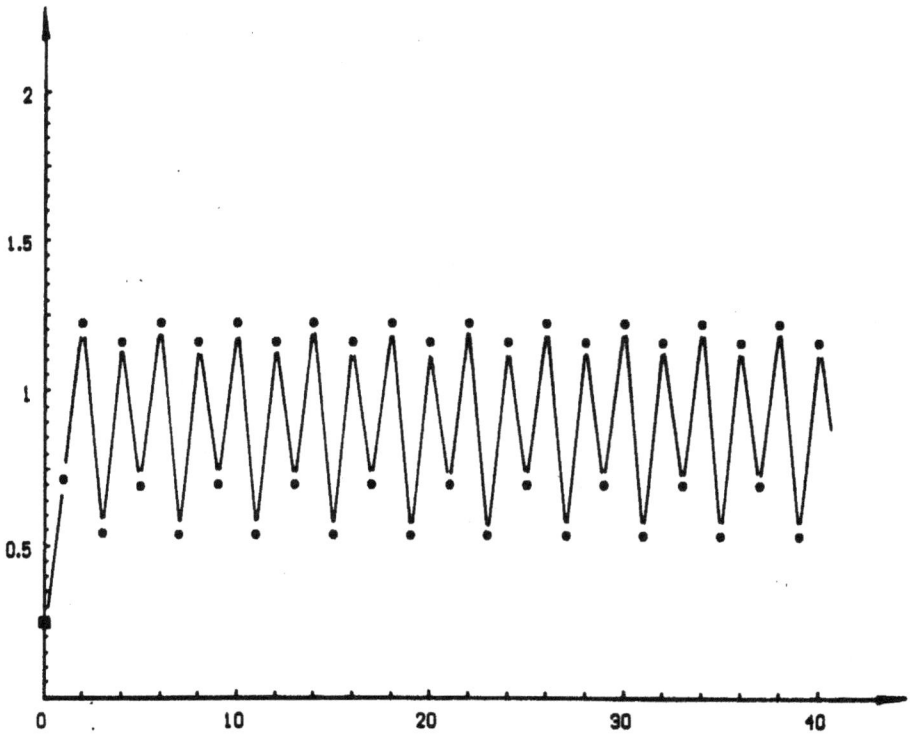

Fig. 12 Quadratic DDS with $\varepsilon = 2.5, \gamma = 1$

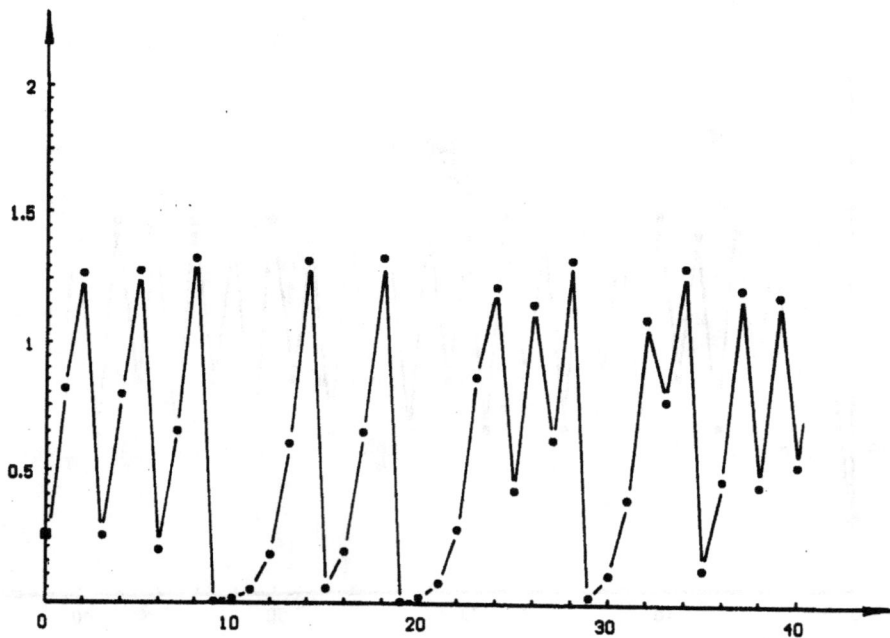

Fig. 13 Quadratic DDS with $\varepsilon = 3, \gamma = 1$

$$| D_x f(f) | = | D_f f | | D_x f | < 1$$

with derivatives computed in the neighbourhood of $x_3{}^*$ & $x_4{}^*$. If so, the periodic solution is stable in the sense that small perturbations from the equilibrium values bring x_k back to them.

Thus $0 < \varepsilon < 2$ enables quadratic DDS to have stable non-zero equilibrium values and $\varepsilon > 2$ to contain stable 2-point (or higher-point) non-zero periodic solutions.

The value $\varepsilon = 2$ is called a *bifurcation point* of the quadratic DDS since for $\varepsilon = 2+\delta$ with $\delta > 0$ as small as we like, the number of equilibrium values doubles and the dynamical behaviour of the DDS changes radically.

As an example, let us consider

$$x_{k+1} = x_k + 2.5 [1 - x_k] x_k$$

of Fig. 12 with $\varepsilon = 2.5$ and $\gamma = 1$ and a stable 4-point periodic solution.

The equilibrium value $x_2{}^* = 1$ is unstable and the 2-point periodic solution is
$$x_3{}^* = 0.6 \qquad \text{and} \qquad x_4{}^* = 1.2$$
from the previous 2-point periodic solution`s formula with

$$| (D_f f)_{f = x4*} | | (D_x f)_{x = x3*} | = 1.25 .$$

So a 2-point periodic solution exists but it is unstable too. Only higher-point stable periodic solutions may exist as it is the case.

Unlike the previous linear DDS, no general solution of quadratic DDS is known to exist. However, for discrete values of $\varepsilon > 0$, closed-form solutions can still be found.

Later, we show that the quadratic DDS of Fig.10 with $\varepsilon = 1$ (and $\gamma = 1$) has the closed-form solution, x_0 given,

$$x_k = \gamma^{-1} [1 - [1 - \gamma x_0]^{2^k}] \qquad\qquad \text{and } 2^k \equiv 2^k$$

for all $\gamma > 0$ and that the "chaotic" quadratic DDS of Fig.13 with $\varepsilon = 3$ (and $\gamma = 1$) has the closed-form solution

$$x_k = \tfrac{2}{3}\,\gamma^{-1}\,[\,1 - \cos(2^k\,\text{arc cos}\,(1 - \tfrac{3}{2}\,\gamma\,x_0\,))\,]$$

for all $\gamma > 0$, $0 \le x_0 \le \tfrac{4}{3}\,\gamma^{-1}$ and

$$x_k = \tfrac{2}{3}\,\gamma^{-1}\,[\,1 - \cosh(2^k\,\text{arc cosh}\,(1 - \tfrac{3}{2}\,\gamma\,x_0\,))\,]$$

for all $\gamma > 0$ and $x_0 > \tfrac{4}{3}\,\gamma^{-1}$.

13. LOGISTIC CDS

The continuous analogue to the previous DDS is:

$$D_t y = \varepsilon [1 - \gamma y] y$$

with $y_0 = y(t_0)$ given and $\varepsilon, \gamma > 0$ constant.

This CDS is referred in the literature as the *Logistic equation* or the *Verhulst-Pearl equation* since it was introduced in biological studies independently by Verhulst and Pearl (see Pearl [13]) in 1922.

The general solution of this differential equation for $y_0 = y(t_0)$ is,

$$y = y_0 \,/\, [\gamma\, y_0 + [1 - \gamma\, y_0]\, e^{-\varepsilon[t-t_0]}]$$

which is a "smooth" function of time $t \geq t_0$ for all values of $\varepsilon, \gamma > 0$.

Since $\lim_{t=\infty} e^{-\varepsilon[t-t_0]} = 0$, we always have $\lim_{t=\infty} y = \gamma^{-1}$ and the solution tends asymptotically to the stable equilibrium value $y^* = \gamma^{-1}$ independently of the values of ε and y_0.

Note that this solution can also be written as

$$y = \gamma^{-1} \,/\, [1 - [1 - [\gamma\, y_0]^{-1}]\, e^{-\varepsilon[t-t_0]}]$$

and this is the sum of the geometric series

$$y = a\, [1 + b\, e^{-\varepsilon[t-t_0]} + b^2\, e^{-2\varepsilon[t-t_0]} + b^3\, e^{-3\varepsilon[t-t_0]} + \ldots]$$

$$= a \,/\, [1 - b\, e^{-\varepsilon[t-t_0]}]$$

with
$$a = \gamma^{-1}$$
$$b = 1 - [\gamma\, y_0]^{-1} < 1,$$

i.e. the solution of the logistic CDS is a sum of an infinite number of exponentials $b^k\, e^{-k\varepsilon[t-t_0]}$ with $\lim_{k=\infty} b^k\, e^{-k\varepsilon[t-t_0]} = 0$.

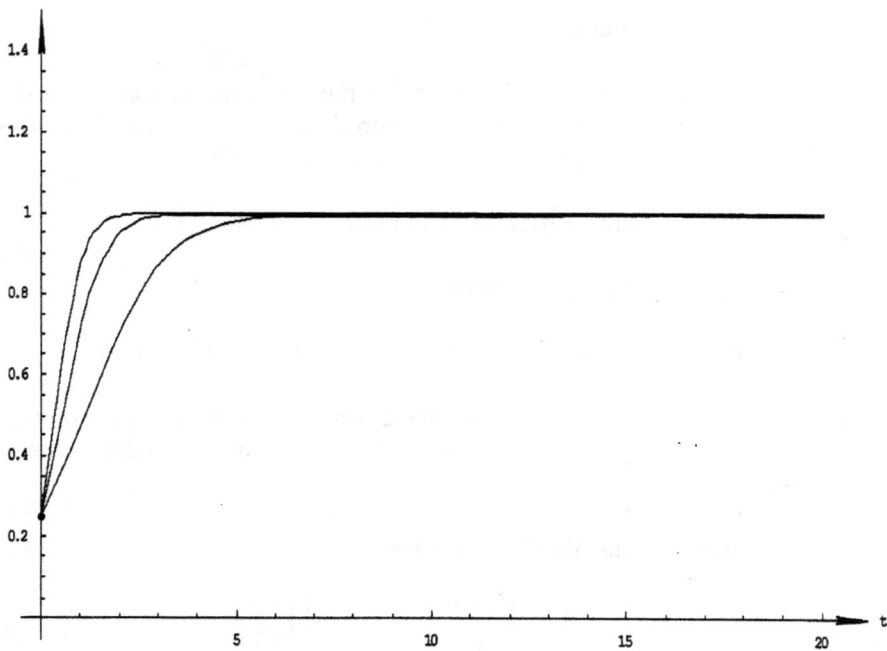

Fig. 14 Logistic CDS with $\varepsilon = 1, 2, 3, \gamma = 1$

Fig. 14 shows 3 solutions of logistic CDS with the same $y_0 = 0.25$ and $\gamma = 1$ but different values for $\varepsilon = 1, 2, 3$. All converge without oscillations to $y^* = 1$ not like the discrete counterparts in Fig. 10, 11, 12, 13.

Sometimes it is important to know when the "speed of growth" of the logistic CDS reaches its maximum since, ultimately, it tends to zero as t tends to infinity.

To find it, we need the zero of

$$D_t^2 \, y \ = \ \varepsilon \, [1 - 2 \, \gamma \, y] \, D_t \, y \qquad \text{which is} \qquad y = \tfrac{1}{2} \, \gamma^{-1}.$$

It can easily be shown that $y = \tfrac{1}{2} \, \gamma^{-1}$ corresponds to a maximum for $D_t \, y$ which is exactly half way between the limiting values for y, $y = 0$ and $y = \gamma^{-1}$ ($0 < y_0 < \gamma^{-1}$).

So, $\max D_t \, y \ = \ \tfrac{1}{4} \, \varepsilon \, \gamma^{-1}$ for $y = \tfrac{1}{2} \, \gamma^{-1}$.

More important, the maximum "speed of growth" is directly proportional to the value of ε and inverse proportional to γ.

This analytical behaviour agrees favourably with the empirical evidence of biological systems where the speed of growth is usually slow to start, gets to a maximum and, then, decreases to zero.

14. SOLUTION OF LOGISTIC CDS

From

$$D_t \, y \; = \; \varepsilon \; [1 - \gamma \, y] \; y$$

and $\varepsilon, \gamma > 0$ constant, we start by separating variables,

$$[\, y \, [1 - \gamma \, y] \,]^{-1} \; dy \; = \; \varepsilon \, dt$$

and splitting the fraction

$$[\, \gamma \, / \, [\gamma \, y] \; + \gamma \, / \, [1 - \gamma \, y] \,] \, dy \; = \; \varepsilon \, dt \, .$$

Integrating

$$\ln \, (\, \gamma \, y \, / \, [1 - \gamma \, y] \,) \; = \; \varepsilon \, t + K^{\#}$$

or

$$[1 - \gamma \, y] \, / \, [\gamma \, y] \; = \; K \, e^{- \varepsilon t}$$

and

$$y \; = \; 1 \, / \, [\gamma + \gamma \, K \, e^{- \varepsilon t}]$$

with

$$K \; = \; [\, 1 \, / \, [\gamma \, y_0] \; - 1] \, e^{\varepsilon t_0} \, .$$

Replacing,

$$\mathbf{y} \; = \; \mathbf{y_0} \; / \; [\, \boldsymbol{\gamma} \, \mathbf{y_0} \; + [1 - \boldsymbol{\gamma} \, \mathbf{y_0}] \, \mathbf{e}^{- \varepsilon [t - t_0]} \,]$$

with a given y_0 and we have the general solution as previously stated.

15. HARVESTING / FISHING CDS

This is the situation when we want to control the growth of a given population by removing continually or periodically some of its own individuals. Obviously, care should be taken with the timing and with the amounts involved, otherwise extinction of the whole population may happen which is not what we usually want.

Here, we shall consider two possible strategies :

Strategy **I** : Fixed/constant harvesting/fishing policy;
Strategy **II** : Proportional harvesting/fishing policy.

Constant harvesting / fishing strategy

In this strategy we shall consider a modified logistic CDS given by :

$$D_t y = \varepsilon [1 - \gamma y] y - \mathbf{H}$$

with $y_0 = y(t_0)$ given and $\varepsilon, \gamma, \mathbf{H} > 0$ constant. The $\mathbf{H}$ represents the constant harvesting/fishing effort.

In a harvesting/fishing context, we usually want to know how big $\mathbf{H}$ can be in a sustainable harvesting/fishing effort.

The first surprise is that there is no simple, algebraicly speaking, general solution for this differential equation. So, to study the long-term behaviour of the solution, we start by finding its equilibrium values.

The equilibrium values $y_1{}^*$ and $y_2{}^*$, are the zeros of the differential equation

$$D_t y = 0 \qquad \rightarrow \qquad \varepsilon \gamma y^2 - \varepsilon y + \mathbf{H} = 0$$

i.e.

$$y_{1,2}{}^* = \tfrac{1}{2} \gamma^{-1} \pm \tfrac{1}{2} \gamma^{-1} [1 - 4 \gamma \varepsilon^{-1} \mathbf{H}]^{\frac{1}{2}} .$$

Note that for $\mathbf{H} = 0$, we have $y_1{}^* = 0$ and $y_2{}^* = \gamma^{-1}$ as in the original logistic equation.

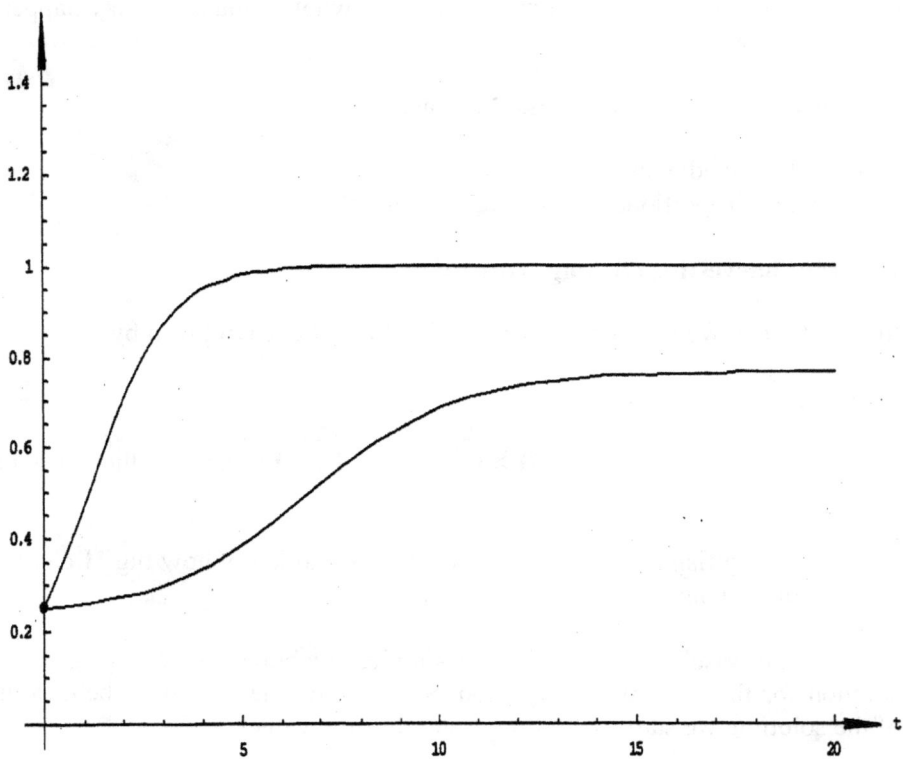

Fig. 15 Constant harvesting/fishing CDS with $\varepsilon = 1, \gamma = 1, \mathbf{H = 0.18}$

For real solutions, we must have,

$$1 - 4\gamma\varepsilon^{-1}\mathbf{H} \geq 0 \qquad \rightarrow \qquad \mathbf{H} \leq \tfrac{1}{4}\,\varepsilon\gamma^{-1}.$$

Not surprisingly, the value $\mathbf{H} = \tfrac{1}{4}\varepsilon\gamma^{-1}$ is called the *maximum sustainable yield* of the logistic CDS.

Replacing it above, we have the non-zero equilibrium value $y_2* = \tfrac{1}{2}\,\gamma^{-1}$ which is half of the non-zero equilibrium value γ^{-1} of the original CDS with $\mathbf{H} = 0$.

Fig. 15 shows a comparison of solutions of two logistic-like CDS with common values for $y_0 = 0.25$, $\varepsilon = \gamma = 1$ and $\mathbf{H} = 0.18$ (bottom curve) and $\mathbf{H} = 0$ (top curve as a reference). Both are smooth curves tending to different equilibrium values. They were obtained using standard numerical integration techniques.

Note that the choice of $\mathbf{H}$ is critical. In fact, for our initial value $y_0 = 0.25$, if we had chosen $\mathbf{H} = 0.2$ (*maximum sustainable yield* is $\mathbf{H} = 0.25$), the whole population would collapse.

Clearly, in practice, strategy **I** is potentially risky since it may bring to extinction the whole population if the population size is not right when we start harvesting/fishing or any of the presumed constant parameters ε, γ change due to unforeseeable circumstances.

Proportional harvesting/fishing strategy

Strategy **II** is, in practice, more difficult to implement since the population size at any given time may not be known precisely or the population unevenly distributed in space.

The modified logistic CDS is given by :

$$D_t\, y \;=\; \varepsilon\,[1 - \gamma\, y]\, y \;-\; \mu\, y$$

with $y_0 = y(t_0)$ given and ε, γ, $\mu > 0$ constant. The μy represents the proportional harvesting/fishing effort. It can be re-written as:

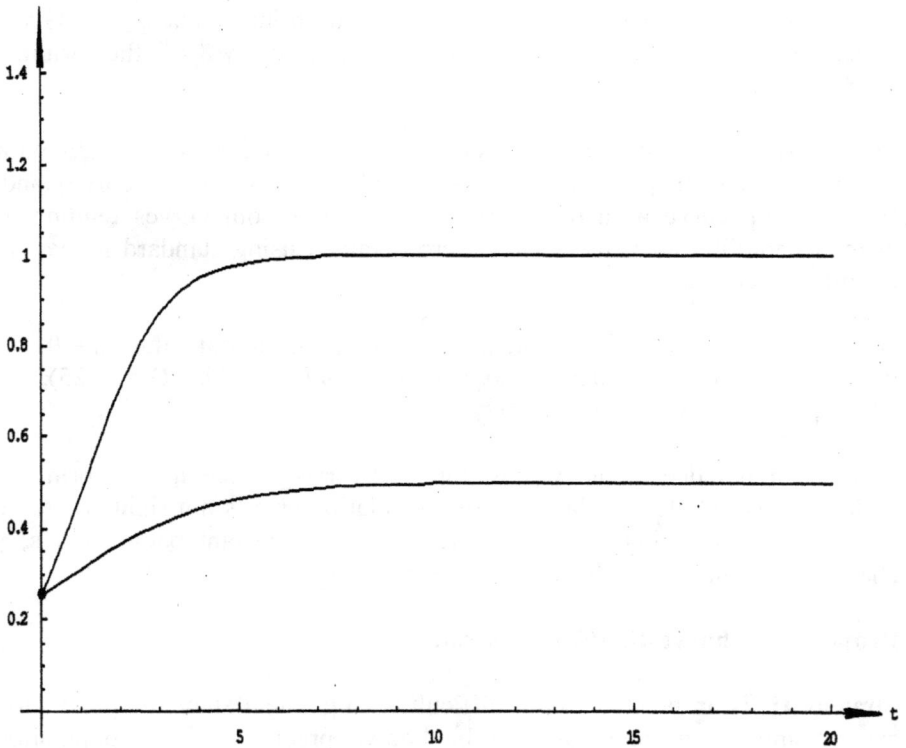

Fig. 16 Proportional harvesting/fishing CDS with $\varepsilon = 1, \gamma = 1, \boldsymbol{\mu = 0.5}$

$$D_t\, y \;=\; \varepsilon\ [1 - \mu/\varepsilon - \gamma\, y]\ y \ .$$

The general solution of this differential equation for a given y_0, is :

$$y \;=\; [1 - \mu/\varepsilon]\ y_0 \ / \ [\gamma\, y_0 \;+\; [1 - \mu/\varepsilon - \gamma\, y_0]\ e^{-[\varepsilon - \mu][t - t_0]}] \ .$$

For $\mu = 0$, it reduces to the one of the original logistic CDS

$$y \;=\; y_0 \ / \ [\gamma\, y_0 \;+\; [1 - \gamma\, y_0]\ e^{-\varepsilon[t - t_0]}] \ .$$

We always have,

$$\lim{}_{t=\infty}\, y \;=\; [1 - \mu/\varepsilon]\, /\, \gamma \qquad \rightarrow \qquad y^* = [1 - \mu/\varepsilon]\, /\, \gamma$$

and the solution tends asymptotically to an equilibrium value that depends not only on the "attrition" coefficient γ but also on the other two parameters ε, μ. Since $\mu > 0$, this harvesting/fishing equilibrium value is always smaller than the one of the original logistic CDS, $y^* = \gamma^{-1}$.

Note that for a sustainable harvesting / fishing effort we need:

$$y^* > 0 \qquad \rightarrow \qquad 1 - \mu/\varepsilon > 0 \qquad \rightarrow \qquad \mu < \varepsilon \ .$$

Fig. 16 shows the solutions of two logistic CDS with common values for $y_0 = 0.25$, $\varepsilon = \gamma = 1$ and $\mu = 0.5$ (bottom curve) and $\mu = 0$ (top curve as a reference). Both solutions are similar in behaviour tending smoothly to different equilibrium values $y^* = 0.5$ and $y^* = 1$ respectively.

Note that the equilibrium value $y^* = 0.5$ above is the same as the equilibrium value of the constant harvesting/fishing strategy at *maximum sustainable yield* for $\gamma = 1$.

Conclusion

So, both strategies allow us to obtain the same non-zero equilibrium value y^* based on the *maximum sustainable yield* of strategy I. The difference being that strategy **II** allows us to obtain it without the risk of population extinction. However, strategy **II** is, in practice, more difficult to implement.

16. HARVESTING / FISHING DDS

Constant harvesting/fishing strategy

For constant harvesting/fishing quadratic DDS, we take :

$$x_{k+1} = x_k + \varepsilon\,[1 - \gamma\,x_k]\,x_k - \mathbf{H}$$

with x_0 given and ε, γ, $\mathbf{H} > 0$ constant. Again, $\mathbf{H}$ represents the constant harvesting/fishing effort.

The equilibrium values x^* and the constraints on the values for $\mathbf{H}$ are essentially the same as the ones of the previous harvesting/fishing CDS analogue when $\varepsilon < 2$.

Fig. 17 shows a comparison for $\varepsilon < 2$ of two quadratic DDS solutions with common values for $x_0 = 0.25$, $\varepsilon = \gamma = 1$ and $\mathbf{H} = 0.18$ (bottom) and $\mathbf{H} = 0$ (top as a reference) and there is great similarity between constant harvesting/fishing discrete and continuous dynamical solutions.

Now we look into the effect of harvesting/fishing on highly oscillating quadratic DDS (not available in logistic CDS) and we start from the "chaotic" DDS of Fig. 13 with $\varepsilon = 3$.

We compare, in Fig.18, two quadratic DDS with common values for $x_0 = 0.25$, $\varepsilon = 3$, $\gamma = 1$, one with $\mathbf{H} = 0.5$ (solid line), the other with $\mathbf{H} = 0$ (broken line) as a reference.

We find that constant harvesting/fishing has a profound stabilizing effect on the oscillating behaviour of the quadratic solution, delaying the *bifurcation points* for values of $\varepsilon > 3$ and producing a stable equilibrium value solution.

We emphasize that the value chosen for $\mathbf{H}$ is somehow critical. A smaller one produces stable periodic solutions and a greater one may, suddenly, bring the whole x_k population to extinction.

Proportional harvesting/fishing strategy

We take the proportional harvesting/fishing DDS to be given by :

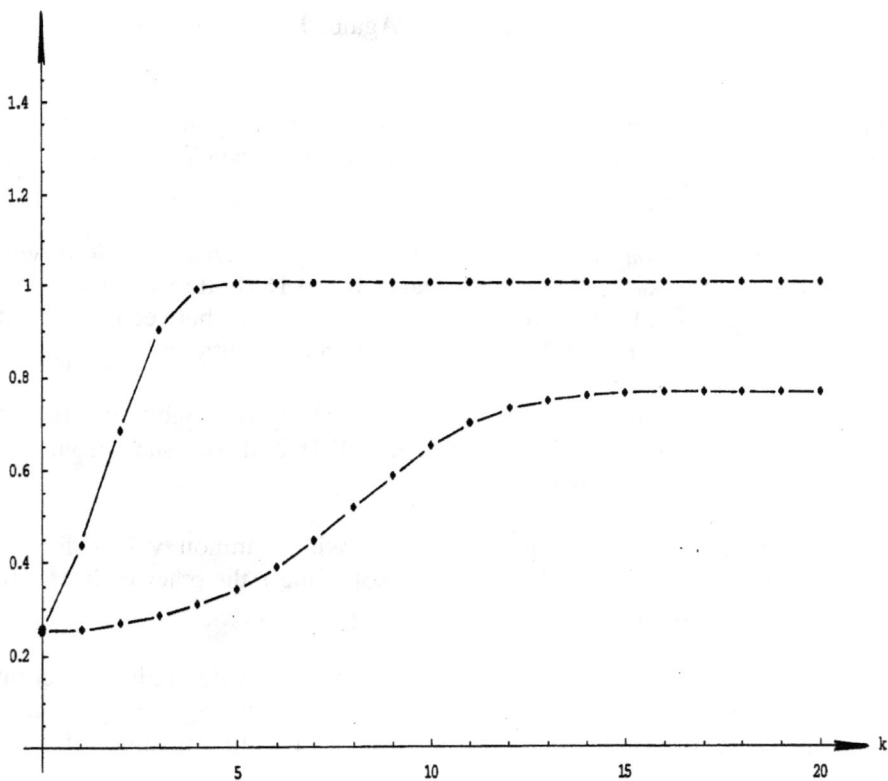

Fig. 17 Constant harvesting/fishing DDS with $\varepsilon = 1, \gamma = 1, \mathbf{H = 0.18}$

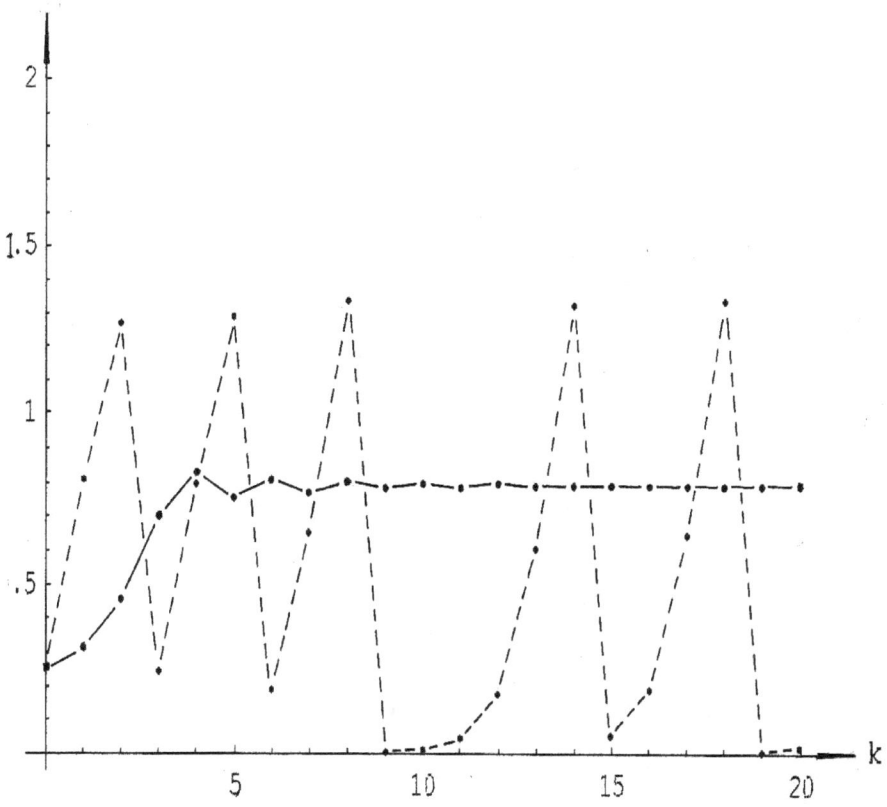

Fig 18 Constant harvesting/fishing DDS with $\varepsilon = 3$, $\gamma = 1$, **H = 0.5**

or

$$X_{k+1} = X_k + \varepsilon [1 - \gamma X_k] X_k - \mu X_k$$

$$X_{k+1} = X_k + \varepsilon [1 - \mu /\varepsilon - \gamma X_k] X_k$$

with x_0 given and $\varepsilon, \gamma, \mu > 0$ constant. Again, μx_k represents the proportional harvesting/fishing effort.

In the absence of a known general solution, we look for the equilibrium values x^* and discuss their stability to have an idea of the long-term behaviour of the solution.

The equilibrium values x^* are the roots of

$$[1 - \mu /\varepsilon - \gamma x] x = 0 \quad \rightarrow \quad x^* = 0 \quad \text{and} \quad x^* = [1 - \mu /\varepsilon] / \gamma$$

and they are the same as the equilibrium values of the continuous counterpart. From

$$f(x) = x + \varepsilon [1 - \mu /\varepsilon - \gamma x] x ,$$

$$(D_x f)_{x^* = 0} = 1 + \varepsilon - \mu > 1$$

and the equilibrium value $x^* = 0$ is unstable since μ must be $\mu < \varepsilon$ for growth to exist in the DDS.

On the contrary,

$$| (D_x f)_{x^* \neq 0} | = | 1 - \varepsilon + \mu | < 1 \qquad \qquad \text{for } 0 < \varepsilon - \mu < 2$$

and stable non-zero equilibrium values exist. Note that here the usual constraint on ε is not $\varepsilon < 2$, but $\varepsilon < 2 + \mu$ and stable equilibrium values may exist for $\varepsilon > 2$,.

Fig.19 shows the solutions of two quadratic DDS with common values for $x_0 = 0.25$, $\varepsilon = \gamma = 1$ and $\mu = 0.5$ (bottom) and $\mu = 0$ (top as a reference). Again for moderate values of the growth coefficient (i.e. $\varepsilon < 2 + \mu$) there is great similarity between proportional harvesting/fishing discrete and continuous dynamical solutions.

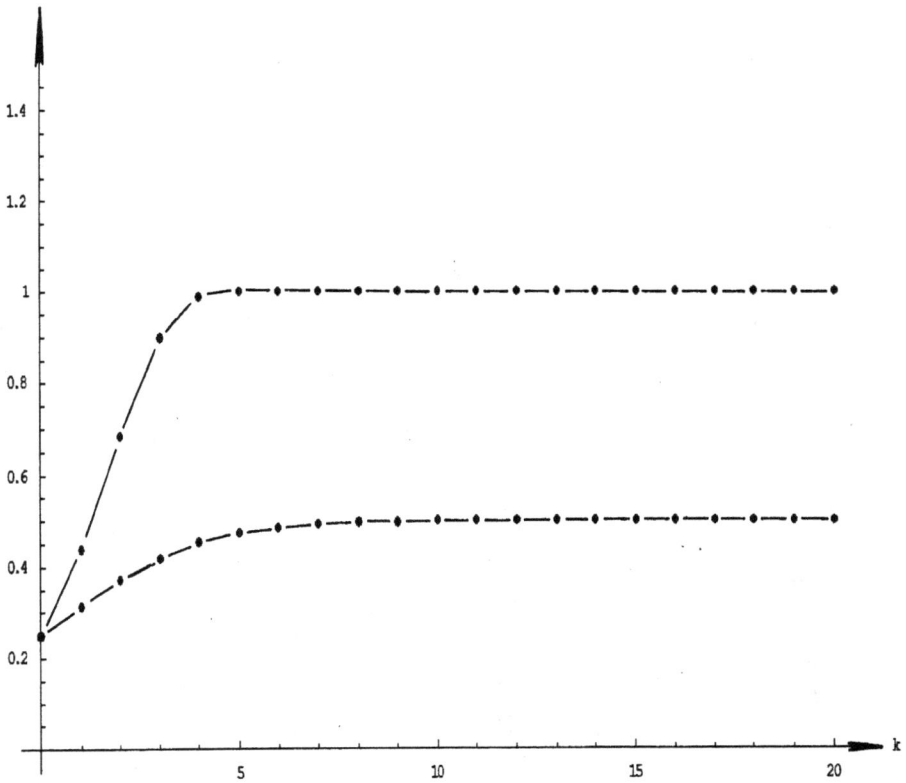

Fig. 19 Proportional harvesting/fishing DDS with $\varepsilon = 1$, $\gamma = 1$, $\mu = 0.5$

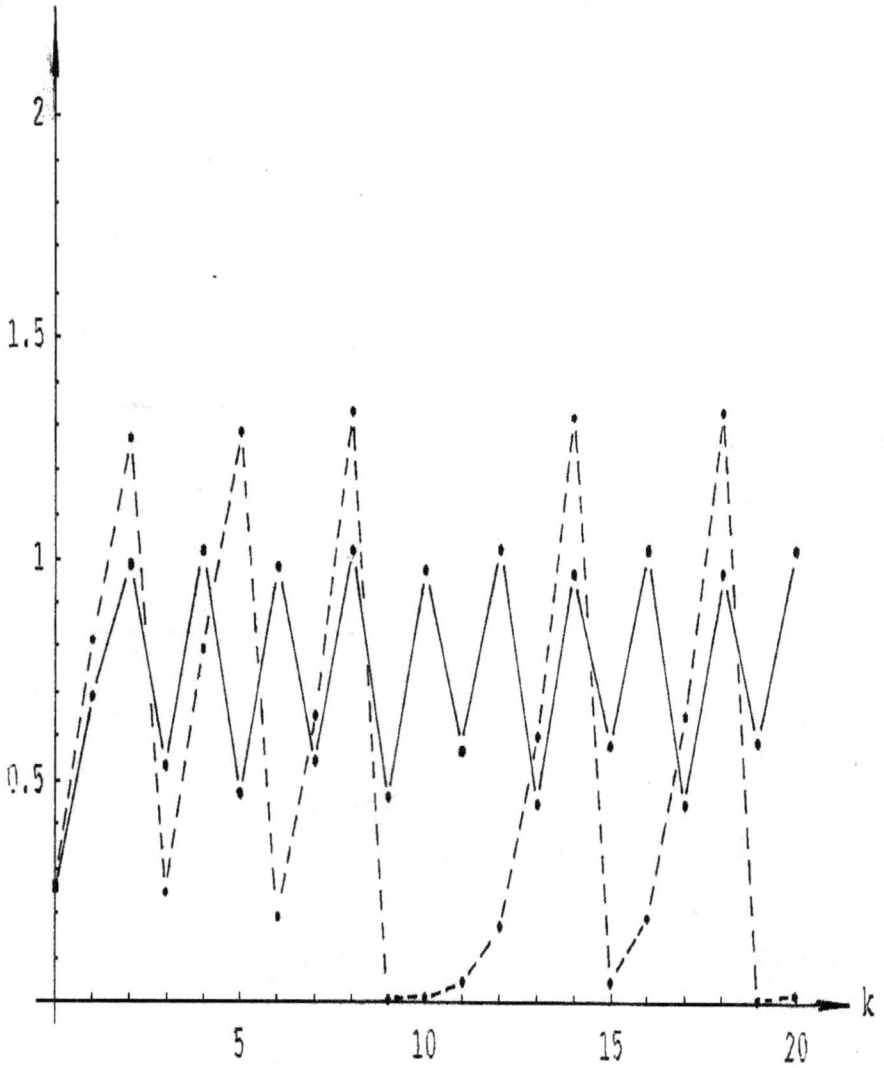

Fig. 20 Proportional harvesting/fishing DDS with $\varepsilon = 3$, $\gamma = 1$, $\mu = 0.5$

Following what we did for the constant harvesting/fishing strategy on highly oscillating quadratic DDS, we compare, in Fig. 20, quadratic DDS with and without proportional harvesting/fishing effort, both with $x_0 = 0.25$, $\varepsilon = 3$, $\gamma = 1$ and $\mu = 0.5$ (solid line) and $\mu = 0$ (broken line) as a reference.

We find that for the same level of harvesting/fishing effort, a proportional harvesting/fishing action produces a less efficient stabilizing effect i.e. a stable periodic 4-point solution (as in Fig. 12) rather than a stable equilibrium value solution of the constant harvesting/fishing one of Fig. 18. The *bifurcation points* change too but the choice of μ is not as critical as **H** is.

17. QUADRATIC CLOSED–FORM SOLUTIONS

As mentioned earlier, no general solution is known for quadratic DDS although closed-form solutions may be found for some.

We start by studying non-periodic non-chaotic type of closed-form solutions of a sub-set of quadratic DDS. Thus, we consider, quadratic DDS in a general algebraic form:

$$x_{k+1} = A [x_k]^2 + 2 B x_k + C$$

with A, B, C constants and x_0 given.

As previously, we shall guess an expression for the closed-form solution where some constants are going to be determined at a later stage.

So we shall try, as a possible solution, an expression of the type:

$$x_k = \beta e^{-\chi_k} + \gamma \qquad\qquad \text{with} \quad \chi_k = \theta_0 2^k$$

with θ_0 known and β, γ constants to be determined later.
Replacing it into the DDS above, we have,

$$x_{k+1} = A \beta^2 e^{-2\chi_k} + 2 \beta [A \gamma + B] e^{-\chi_k} + A \gamma^2 + 2 B \gamma + C .$$

Since

$$x_{k+1} = \beta e^{-2\chi_k} + \gamma ,$$

we obtain, equating coefficients of corresponding exponentials,

$$\beta = A^{-1}$$
$$\gamma = -A^{-1}B$$
$$C = -A^{-1}B [B - 1] .$$

Thus, for all A or B, the quadratic DDS

$$x_{k+1} = A [x_k]^2 + 2 B x_k - A^{-1}B [B - 1]$$

have the closed solution

$$x_k = A^{-1} e^{-\chi_k} - B/A \qquad\qquad \text{with} \quad \chi_k = \theta_0\, 2^k .$$

To express θ_0 as a function of x_0 make $k = 0$ in the solution and invert it

$$- \theta_0 = \ln(A x_0 + B) \qquad\qquad \text{whenever} \quad A x_0 + B > 0 .$$

Replacing θ_0 into the solution, we have

$$x_k = A^{-1} e^{\ln(A x_0 + B)\,2^k} - B/A$$

or

$$x_k = A^{-1} [A x_0 + B]^{2^k} - B/A \qquad\qquad \text{with} \quad 2\hat{\ }k \equiv 2^k$$

which is a closed-form exponential solution for A and B, $A x_0 + B > 0$.

Note that for $|A x_0 + B| < 1$ a stable equilibrium exists since

$$x^* = \lim_{k=\infty} x_k = - B/A .$$

Now, let us compare this result with our original quadratic DDS written as a function of the "growth" and "attrition" coefficients $\varepsilon, \gamma > 0$. We have

$$x_{k+1} = - \varepsilon \gamma [x_k]^2 + [\varepsilon + 1] x_k + 0$$

with
$$A^{-1}B[B - 1] = 0 \quad \rightarrow \quad B = 1$$
$$\tfrac{1}{2}[\varepsilon + 1] = B = 1 \quad \rightarrow \quad \varepsilon = 1$$
$$- \varepsilon \gamma = A \quad\qquad \rightarrow \quad A = -\gamma .$$

In Fig.10, we have the plot of a quadratic DDS with $\varepsilon = 1$ which gives $B = 1$ and $A = -\gamma$. So, this quadratic DDS has the closed solution

$$x_k = \gamma^{-1} [1 - [1 - \gamma x_0]^{2^k}] \qquad\qquad \text{with} \quad 2\hat{\ }k \equiv 2^k$$

for unrestricted $\gamma > 0$, x_0 given. A stable equilibrium exists for $|1 - \gamma x_0| < 1$ with $x^* = \gamma^{-1}$. In Fig. 10, $\gamma = 1$ and $x^* = 1$.

In this closed solution, the "attrition" coefficient $\gamma > 0$ acts as an overall brake on the growth of the solution.

Chaotic type closed-form solutions

Let us consider again quadratic DDS in a general form:

$$x_{k+1} = A\,[x_k]^2 + 2\,B\,x_k + C$$

with A, B, C constants and x_0 given and let us try now, as a possible solution, an expression of the type:

$$x_k = \alpha\,e^{i\chi_k} + \beta\,e^{-i\chi_k} + \gamma \qquad\qquad \text{with } \chi_k = \theta_0\,2^k$$

with θ_0 known and α, β, γ constants to be determined later ($i^2 = -1$).

Replacing it into the quadratic equation DDS above, we have,

$$x_{k+1} = A\,\alpha^2\,e^{i2\chi_k} + A\,\beta^2\,e^{-i2\chi_k} + 2\,\alpha\,[A\,\gamma + B]\,e^{i\chi_k} +$$
$$+ 2\,\beta\,[A\,\gamma + B]\,e^{-i\chi_k} + 2\,A\,\alpha\,\beta + A\,\gamma^2 + 2\,B\,\gamma + C .$$

Since

$$x_{k+1} = \alpha\,e^{i2\chi_k} + \beta\,e^{-i2\chi_k} + \gamma ,$$

equating coefficients of corresponding exponentials, we obtain,

$$\alpha = \beta = A^{-1}$$
$$\gamma = -A^{-1}B$$
$$C = A^{-1}[B^2 - B - 2] .$$

Then, for all A or B the quadratic DDS,

$$x_{k+1} = A\,[x_k]^2 + 2\,B\,x_k + [B^2 - B - 2]\,/\,A$$

has the closed solution

$$x_k = A^{-1}\,[e^{i\chi_k} + e^{-i\chi_k}] - B\,/A \qquad\qquad \text{with } \chi_k = \theta_0\,2^k .$$

By Euler's formula

$$e^{i\chi_k} = \cos\chi_k + i\,\sin\chi_k$$

and the solution becomes

$$x_k = 2 A^{-1} \cos \chi_k - B/A$$

or

$$x_k = 2 A^{-1} \cos(\theta_0 2^k) - B/A .$$

To express θ_0 as a function of x_0, we invert the solution with $k = 0$ to get

$$\theta_0 = \text{arc cos}(\tfrac{1}{2} A x_0 + \tfrac{1}{2} B) \qquad \text{whenever } |A x_0 + B| \le 2 .$$

Replacing θ_0 into the solution, we obtain

$$\mathbf{x_k = 2 A^{-1} \cos(2^k \text{ arc cos}(\tfrac{1}{2} A x_0 + \tfrac{1}{2} B)) - B/A}$$

which is a **periodic** closed solution in the usual form valid for all A or B with $|A x_0 + B| \le 2$. This solution is highly oscillatory and very sensitive regarding x_0 since it can be expressed as

$$\mathbf{x_k = 2 A^{-1} T_{2k}(\tfrac{1}{2} A x_0 + \tfrac{1}{2} B) - B/A}$$

where

$$T_n(x) = \cos(n \text{ arc cos } x)$$

is the Chebyshev polynomial of degree $n = 2^k$ in x, $|x| \le 1$.

For $A > 0$ and $B = 0$, we have $C = -2 A^{-1}$, the quadratic DDS

$$x_{k+1} = A [x_k]^2 - 2 A^{-1}$$

has a simpler closed solution

$$\mathbf{x_k = 2 A^{-1} \cos(2^k \text{ arc cos}(\tfrac{1}{2} A x_0))} \qquad \text{for} \quad |x_0| \le 2 A^{-1} .$$

In Fig. 21, for $A = 2$ and $B = 0$, we have for increastng k, two solutions of

$$x_{k+1} = 2 [x_k]^2 - 1$$

with $x_0 = 0.25, 0.25 - \delta$ and $\delta = 10^{-4}$. Note that the solutions change from highly oscillatory to almost periodic behaviour and then back to an highly oscillatory one. Their apparent randomness is striking.

Let us now consider general quadratic DDS **without** a constant term where "chaotic" solutions can be observed. We need

$$C = B^2 - B - 2 = 0 \quad \rightarrow \quad B = 2 \quad \text{or} \quad B = -1$$

and the corresponding quadratics become:

$$x_{k+1} = A [x_k]^2 + 4 x_k$$

or

$$x_{k+1} = A [x_k]^2 - 2 x_k.$$

For the first one with $B = 2$, the periodic closed solution is

$$x_k = 2 A^{-1} [\cos(2^k \arccos(\tfrac{1}{2} A x_0 + 1)) - 1] \quad \text{for } |x_0| \leq 2 A^{-1}.$$

Let us now compare it with our original quadratic DDS without a constant term written as a function of the "growth" and "attrition" coefficients $\varepsilon, \gamma > 0$

$$x_{k+1} = -\varepsilon \gamma [x_k]^2 + [\varepsilon + 1] x_k.$$

For them to be periodic, we need

$$B = 2 \rightarrow \varepsilon + 1 = 4 \quad \text{or} \quad B = -1 \rightarrow \varepsilon + 1 = -2 \quad \text{or} \quad \varepsilon = 3$$

since $\varepsilon > 0$. So, the quadratic DDS in Fig.13 with $\varepsilon = 3$ and $B = 2$ and $A = -3\gamma$, has the periodic closed solution

$$x_k = \tfrac{2}{3} \gamma^{-1} [1 - \cos(2^k \arccos(1 - \tfrac{3}{2}\gamma x_0))]$$

for unrestricted $\gamma > 0$ and $x_0 \leq \tfrac{4}{3}\gamma^{-1}$, x_0 given. In Fig. 13, $\gamma = 1$.

The closed solution oscillates non-stop between $\min(x_k) = 0$ and $\max(x_k) = \tfrac{4}{3}\gamma^{-1}$ and the amplitude of oscillations decreases for increasing γ.

Note that the existence of periodic solutions in quadratic DDS depends on the value of the growth coefficient ε and on the initial value x_0 but it is independent of γ. The restriction on x_0 for the existence of "chaotic" solutions is not usually acknowledged.

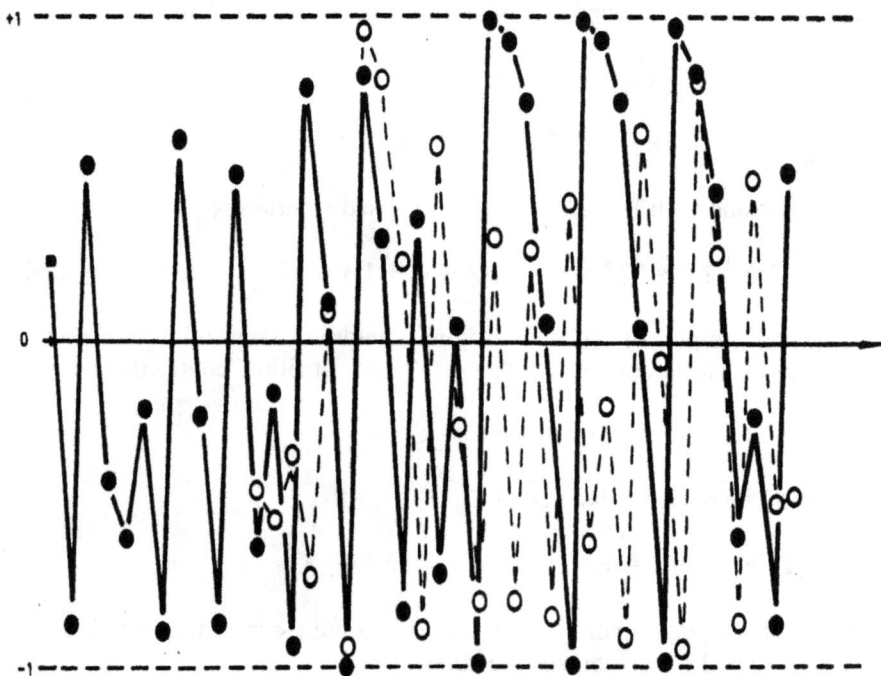

Fig. 21 Quadratic DDS with "chaotic" solutions $x_{k+1} = 2 [x_k]^2 - 1$
$x_0 = 0.2499$ & $x_0 = 0.2500$

Unbounded closed-form solutions

For values of x_0 in $|A x_0 + B| > 2$, we can try a new solution of the form

$$x_k = \alpha\, e^{\chi_k} + \beta\, e^{-\chi_k} + \gamma \qquad\qquad \text{with} \quad \chi_k = \theta_0\, 2^k$$

with θ_0 known and α, β, γ constants to be determined later.

Using the same method as previously, it can be shown, that the closed solution

$$\mathbf{x}_k = 2\, A^{-1} \cosh(2^k \text{ arc cosh} (\tfrac{1}{2} A x_0 + \tfrac{1}{2} B)) - B/A$$

is obtained for the quadratic DDS

$$x_{k+1} = A\, [\mathbf{x}_k]^2 + 2\, B\, \mathbf{x}_k + [B^2 - B - 2]/A, \qquad |A x_0 + B| > 2.$$

These solutions are no longer bounded but grow exponentially.

As an example, let us consider the previous quadratic with $B = 0$, $C = -2\, A^{-1}$

$$x_{k+1} = A\, [x_k]^2 - 2\, A^{-1}.$$

It has the unbounded closed solution

$$\mathbf{x}_k = 2\, A^{-1} \cosh(2^k \text{ arc cosh}(\tfrac{1}{2} A x_0)) \qquad\qquad \text{for} \quad |x_0| > 2\, A^{-1}.$$

Also
$$x_{k+1} = A\, [x_k]^2 + 4\, x_k$$

with $A = -3\, \gamma$ and $B = 2$, $(A^{-1}[B^2 - B - 2] = 0)$ has, in the ε, $\gamma > 0$ notation, the unbounded closed solution

$$\mathbf{x}_k = \tfrac{2}{3}\, \gamma^{-1} [\, 1 - \cosh(2^k \text{ arc cosh}(1 - \tfrac{3}{2}\, \gamma\, x_0))\,] \qquad \text{for} \quad x_0 > \tfrac{4}{3}\, \gamma^{-1}.$$

18. STABILITY OF HIGHLY-OSCILLATORY SOLUTIONS

To study the sensitivity of these closed-form solutions to the initial value x_0, we take the previous closed solution

$$x_k = 2 A^{-1} \cos (2^k \arccos (\tfrac{1}{2} A x_0 + \tfrac{1}{2} B)) - B/A$$

with $| A x_0 + B | \leq 2$ and the error formula:

$$| \Delta x_k | = | (D_{x0} \; x_k)_{x0=x\eta} | \; | \Delta x_0 | , \qquad x_\eta \simeq x_0 .$$

$$| D_{x0} \; x_k | = 2^k \; \frac{| \sin (2^k \arccos (\tfrac{1}{2} A x_0 + \tfrac{1}{2} B)) |}{[1 - [\tfrac{1}{2} A x_0 + \tfrac{1}{2} B]^2]^{\tfrac{1}{2}}}$$

and the error $| \Delta x_k |$ is magnified by an exponential 2^k factor in $| D_{x0} \; x_k |$ for minor errors/variations $| \Delta x_0 |$ in the x_0, x_0 within $| A x_0 + B | < 2$. For $| A x_0 + B | = 2$, the error magnification is even worse, since the denominator is zero and the magnification infinite.

So, these closed solutions are truly unstable for all x_0 within $| A x_0 + B | \leq 2$. They are often called "chaotic" since minor variations or errors on their initial x_0 produce major changes in x_k.

To illustrate their erratic behaviour we plotted in Fig. 22, using 20-digit decimal precision, the solutions of

$$x_{k+1} = 2 [x_k]^2 - 1$$

with $x_0 = 0.75, 0.75 - \delta$ and $\delta = 10^{-4}$. For values of $k > 10$ the solutions are so different in character that it is hard to believe that they are from the same relatively simple recurrence relation.

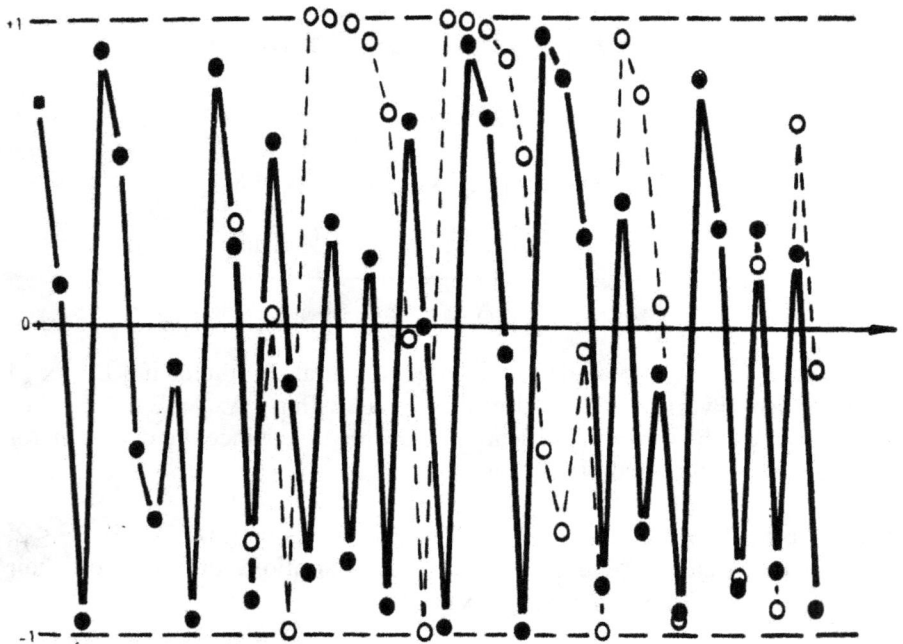

Fig. 22 Instability of "chaotic" solutions with initial conditions
$x_0 = 0.7499$ & $x_0 = 0.7500$

74

19. CLOSED – FORM SOLUTIONS FOR CUBIC DDS

We consider cubic DDS in the general form

$$x_{k+1} = A [x_k]^3 + B [x_k]^2 + C x_k + D$$

with A, B, C, D constant and x_0 given.

We start by guessing the algebraic expression for the solution as:

$$x_k = \alpha e^{i\chi_k} + \beta e^{-i\chi_k} + \gamma \qquad \text{with } \chi_k = \theta_0 3^k$$

with θ_0 known and α, β, γ constants to be determined later ($i^2 = -1$). Replacing it into the cubic DDS above, we have

$$
\begin{aligned}
x_{k+1} = {} & A \alpha^3 e^{i3\chi_k} + B \beta^3 e^{-i3\chi_k} + \\
& + [B \alpha^2 + 3 A \alpha^2 \gamma] e^{i2\chi_k} + [B \beta^2 + 3 A \beta^2 \gamma] e^{-i2\chi_k} + \\
& + [3 A \alpha^2 \beta + 3 A \alpha \gamma^2 + 2 B \alpha \gamma + C \alpha] e^{i\chi_k} + \\
& + [3 A \alpha \beta^2 + 3 A \beta \gamma^2 + 2 B \beta \gamma + C \beta] e^{-i\chi_k} + \\
& + 6 A \alpha \beta \gamma + 2 B \alpha \beta + A \gamma^3 + B \gamma^2 + A \gamma + D .
\end{aligned}
$$

Since

$$x_{k+1} = \alpha e^{i3\chi_k} + \beta e^{-i3\chi_k} + \gamma ,$$

we obtain, equating coefficients of corresponding exponentials,

$$\alpha = \beta = \pm A^{-\frac{1}{2}}, \qquad \gamma = -\tfrac{1}{3} A^{-1} B$$

and

$$C = \tfrac{1}{3} A^{-1} B^2 - 3$$
$$D = \tfrac{1}{3} A^{-1} B [\tfrac{1}{9} A^{-1} B^2 - 4]$$

Then, for all $A > 0$ or B, the cubic DDS

$$x_{k+1} = A [x_k]^3 + B [x_k]^2 + [\tfrac{1}{3}A^{-1} B^2 - 3] x_k + \tfrac{1}{3} A^{-1} B [\tfrac{1}{9} A^{-1} B^2 - 4]$$

has the closed solution

$$x_k = \pm A^{-\frac{1}{2}} [e^{i\chi_k} + e^{-i\chi_k}] - \tfrac{1}{3} B / A \qquad \text{with } \chi_k = \theta_0 3^k$$

and, using Euler's formula,

$$x_k = \pm 2\, A^{-\frac{1}{2}} \cos(\theta_0 3^k) - \tfrac{1}{3} B/A$$

or

$$\mathbf{x_k = \pm 2\, A^{-\frac{1}{2}} \cos(3^k \arccos(\pm \tfrac{1}{2} A^{-\frac{1}{2}} [A\, x_0 + \tfrac{1}{3} B])) - \tfrac{1}{3} B/A}$$

if we replace the value of θ_0 as a function of x_0 and $|A\, x_0 + \tfrac{1}{3} B| \leq 2\, A^{\frac{1}{2}}$. This closed solution is highly oscillatory since it can be written as

$$\mathbf{x_k = \pm 2\, A^{-\frac{1}{2}} T_{3k}(\pm \tfrac{1}{2} A^{-\frac{1}{2}} [A\, x_0 + \tfrac{1}{3} B]) - \tfrac{1}{3} B/A}$$

where

$$T_n(x) = \cos(n \arccos x)$$

is the Chebyshev polynomial of degree $n = 3^k$ in x, $|x| \leq 1$.

If we take $A > 0$, $B = 0$, $C = -3$ and $D = 0$ the cubic DDS

$$x_{k+1} = A\, [x_k]^3 - 3\, x_k$$

has a simpler closed solution

$$\mathbf{x_k = \pm 2\, A^{-\frac{1}{2}} \cos(3^k \arccos(\pm \tfrac{1}{2} A^{\frac{1}{2}} x_0))} \qquad \text{for } |x_0| \leq 2\, A^{-\frac{1}{2}}.$$

However, the cubic DDS

$$x_{k+1} = -4\, [x_k]^3 + 3\, x_k$$

x_0 given, has the real solution:

$$x_k = \sin(3^k \arcsin(x_0)) \qquad \text{for } |x_0| \leq 1.$$

In fact, if we replace x_k by $\sin(z_k)$ with

$$z_k = 3^k \arcsin(x_0)$$

into the cubic DDS, we obtain the trigonometric identity

$$\sin 3 z_k = -4\, [\sin z_k]^3 + 3 \sin z_k \qquad \text{with } z_k \text{ real.}$$

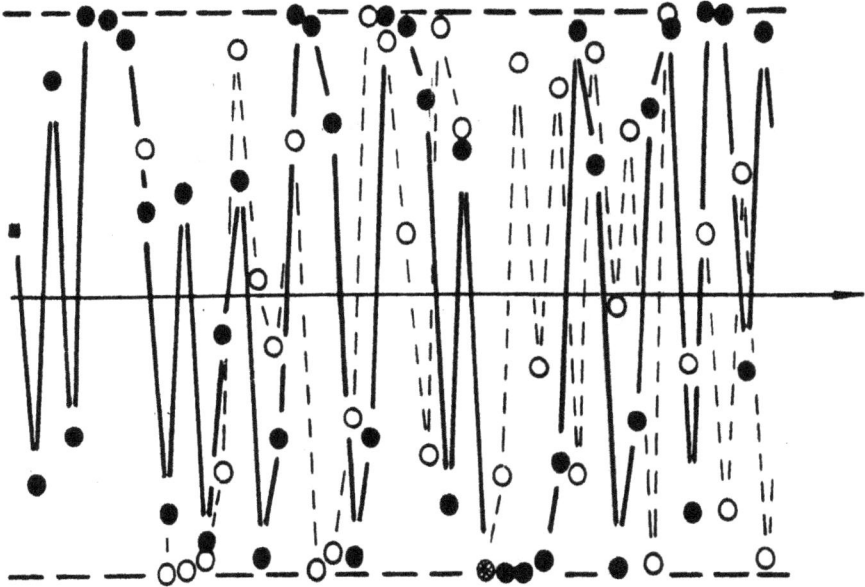

Fig. 23 Cubic DDS with "chaotic" solutions $x_{k+1} = 4\,[x_k]^3 - 3\,x_k$
$x_0 = 0.2499$ & $x_0 = 0.2500$

For $A = 4$, $B = 0$, $C = -3$ and $D = 0$ we have in Fig. 23 two solutions of

$$x_{k+1} = 4 [x_k]^3 - 3 x_k$$

with $x_0 = 0.25$, $0.25 - \delta$ and $\delta = 10^{-4}$. The solutions are again highly oscillatory with great apparent randomness and are very sensitive regarding the initial x_0. It can be shown, as we did for the quadratic solutions, that the error magnifying factor is now 3^k. It is remarkable their randomness.

Unbounded closed-form solutions

Similarly, for values of x_0 in $| A x_0 + \frac{1}{3} B | > 2 A^{\frac{1}{2}}$ we can try

$$x_k = \alpha e^{\chi_k} + \beta e^{-\chi_k} + \gamma \qquad \text{with} \quad \chi_k = \theta_0 3^k$$

with θ_0 known and α, β, γ constants to be determined later.

Using the same method as above, it can be shown, that the closed solution

$$\mathbf{x_k = \pm 2 A^{-\frac{1}{2}} \cosh (3^k \, arc \, cosh (\pm \frac{1}{2} A^{-\frac{1}{2}} [A x_0 + \frac{1}{3} B])) - \frac{1}{3} B / A}$$

is obtained for the cubic DDS

$$x_{k+1} = A [x_k]^3 + B [x_k]^2 + [\frac{1}{3} A^{-1} B^2 - 3] x_k + \frac{1}{3} A^{-1} B [\frac{1}{9} A^{-1} B^2 - 4]$$

when $| A x_0 + \frac{1}{3} B | > 2 A^{\frac{1}{2}}$. The solutions are no longer bounded but grow exponentially.

As an example, the previous cubic DDS with $A > 0$, $B = 0$ and x_0 given,

$$x_{k+1} = A [x_k]^3 - 3 x_k$$

and $| x_0 | > 2 A^{-\frac{1}{2}}$, has the closed solution

$$\mathbf{x_k = \pm 2 A^{-\frac{1}{2}} \cosh (3^k \, arc \, cosh (\pm \frac{1}{2} A^{\frac{1}{2}} x_0)) } \ .$$

For quartics and higher-degree DDS, we can still find closed-form solutions although the algebra gets cumbersome. See Oliveira-Pinto & Adibpour [11] for further study.

Other closed-form solutions

Let us consider again cubic DDS in the general form

$$x_{k+1} = A [x_k]^3 + B [x_k]^2 + C x_k + D$$

with A, B, C, D constant and x_0 given and let us try now,as possible solution,

$$x_k = \beta e^{-\chi_k} + \gamma \qquad\qquad \text{with } \chi_k = \theta_0 3^k$$

with θ_0 known and β, γ constants to be determined later.

Replacing it into the cubic DDS above, we have

$$
\begin{aligned}
x_{k+1} = \ & A \beta^3 e^{-3\chi_k} + [3 A \beta^2 \gamma + B \beta^2] e^{-2\chi_k} + \\
& + [3 A \beta \gamma^2 + 2 B \beta \gamma + C \beta] e^{-\chi_k} + \\
& + A \gamma^3 + B \gamma^2 + A \gamma + D
\end{aligned}
$$

with

$$x_{k+1} = \beta e^{-3\chi_k} + \gamma \ .$$

Equating coefficients of corresponding exponentials,

$$\beta = \pm A^{-\frac{1}{2}}$$
$$\gamma = -\tfrac{1}{3} A^{-1} B$$

and

$$C = \tfrac{1}{3} A^{-1} B^2$$
$$D = \tfrac{1}{3} A^{-1} B [A^{-1} [\tfrac{1}{3}B]^2 - 1] \ .$$

So, for all $A > 0$ or B, the cubic DDS

$$x_{k+1} = A [x_k]^3 + B [x_k]^2 + \tfrac{1}{3}A^{-1} B^2 x_k + \tfrac{1}{3} A^{-1} B [A^{-1} [\tfrac{1}{3}B]^2 - 1].$$

has the closed solution

$$x_k = \pm A^{-\frac{1}{2}} e^{-\chi_k} - \tfrac{1}{3} B /A \qquad\qquad \text{with } \chi_k = \theta_0 3^k \ .$$

To obtain the closed solution as a function of x_0, we invert it and, for $k = 0$,

$$- \theta_0 = \ln (| A^{-\frac{1}{2}} [A x_0 + \tfrac{1}{3} B] |) .$$

Replacing, the solution becomes

$$x_k = \pm A^{-\frac{1}{2}} e^{\ln (A^{\wedge} [-\frac{1}{2}] | A x_0 + \frac{1}{3} B |) 3^{\wedge} k} - \tfrac{1}{3} B / A$$

or

$$x_k = \pm A^{-\frac{1}{2}} [A^{-\frac{1}{2}} [A x_0 + \tfrac{1}{3} B]]^{3^{\wedge} k} - \tfrac{1}{3} B / A \qquad \text{with } 3^{\wedge} k \equiv 3^k$$

which is a closed exponential solution for $A > 0$ and B, with x_0 given .

A stable equilibrium x^* exists when

$$A^{-\frac{1}{2}} | A x_0 + \tfrac{1}{3} B | < 1 \qquad \text{or} \qquad | A x_0 + \tfrac{1}{3} B | < A^{\frac{1}{2}}$$

since

$$x^* = \lim_{k=\infty} x_k = - \tfrac{1}{3} B / A .$$

Note that to have $x^* \neq 0$, we need $B \neq 0$ and $C \neq 0$.

As an example, let us take:

$$x_{k+1} = [x_k]^3 - 3 [x_k]^2 + 3 x_k \qquad \text{with } x_0 \text{ given}$$

where the coefficients fulfil the conditions ($A = 1, B = - 3, C = 3, D = 0$).

The closed exponential solution is

$$x_k = 1 \pm [x_0 - 1]^{3^{\wedge} k} \qquad \text{with } 3^{\wedge} k \equiv 3^k$$

x_0 given. For $0 < x_0 < 2$ we have

$$x^* = \lim_{k=\infty} x_k = 1$$

and a stable equilibrium $x^* = 1$ exists with the solution tending to the equilibrium value very quickly. The equilibrium $x^* = 0$ also exists but it is an unstable one.

20. STABILITY CONDITION FOR 2nd ORDER DDS

For 1st order DDS

$$x_{k+1} = f(x_k), \qquad\qquad \text{with } x_0 \text{ given}$$

with a known equilibrium value x^*, we found that the condition for stability at x^* is $|D_x f| < 1$ for x in the neighbourhood of x^*.

Now, for a DDS of the form :

$$\begin{cases} x_{k+1} = f(x_k, y_k) \\[2em] y_{k+1} = g(x_k, y_k), \qquad\qquad \text{with } x_0 \;\&\; y_0 \text{ given} \end{cases}$$

with known equilibrium values x^* & y^* for x_k and y_k, we shall not consider the derivatives of individual recurrence relations but we shall look instead at the eigenvalues of the determinant of the partial derivatives of the recurrence relations and construct the Jacobian

$$\begin{bmatrix} \partial_x f & \partial_y f \\[2em] \partial_x g & \partial_y g \end{bmatrix}$$

with the partial derivatives computed at the equilibrium points (x^*, y^*) of x_k and y_k respectively.

The so-called eigenvalues are the roots r_1 & r_2 of the determinant

$$\begin{vmatrix} [\partial_x f - r] & \partial_y f \\[2em] \partial_x g & [\partial_y f - r] \end{vmatrix} = 0 .$$

and an equilibrium point is said to be stable, in the sense that

$$|x_{k+1} - x_k| < |x_k - x_{k-1}|$$

and

$$|y_{k+1} - y_k| < |y_k - y_{k-1}|$$

if both eigenvalues r_1 & r_2 are $|r_1| < 1$ **and** $|r_2| < 1$ [or, which is equivalent, $max(|r_1|, |r_2|) < 1$] .

Otherwise, the equilibrium point is said to be unstable unless $|r_1| = 1$ or $|r_2| = 1$ when a more detailed study is needed. In a similar way, we may extend this stability condition to DDS with three or more recurrence relations .

21. LINEAR COUPLED DDS

Let us consider linear DDS as a couple of linear recurrence relations

$$\begin{cases} x_{k+1} = a_1 x_k + b_1 y_k + c_1 \\ \\ y_{k+1} = a_2 x_k + b_2 y_k + c_2 \end{cases}$$

with constant coefficients and x_0 & y_0 given, $k = 0, 1, 2\ldots$.

Before we look for the general solution of this linear DDS we transform these recurrence relations into 2nd order ones with separated variables.

First, from the 1st one, we get

$$y_k = [x_{k+1} - a_1 x_k - c_1] / b_1 .$$

Replacing it into the 2nd one, we have

$$x_{k+2} = [a_1 + b_2] x_{k+1} + [a_2 b_1 - a_1 b_2] x_k + [1 - b_2] c_1 + b_1 c_2$$

or, simply,

$$x_{k+2} = A_1 x_{k+1} + B_1 x_k + C_1 .$$

Likewise, we may eliminate x_k from the 2nd one, to get

$$y_{k+2} = [a_1 + b_2] y_{k+1} + [a_2 b_1 - a_1 b_2] y_k + [1 - a_1] c_2 + a_2 c_1$$

or

$$y_{k+2} = A_2 y_{k+1} + B_2 y_k + C_2 .$$

Thus, $A_1 = A_2 = A$ and $B_1 = B_2 = B$ but $C_1 \neq C_2$ likely.

2nd order homogeneous relations

Let us now consider one of the 2 previous 2nd order recurrence relations as an homogeneous relation, i.e.

$$x_{k+2} = A\, x_{k+1} + B\, x_k$$

with A and B constant and x_0 & x_1 ($x_1 = a_1 x_0 + b_1 y_0$) given.

We start by showing that the solution of this 2nd order recurrence relation has the algebraic form

$$x_k = \alpha\, r^k + \beta\, s^k$$

where r and s are the roots of the quadratic equation, called the *characteristic equation* as in the theory of differential equations,

$$x^2 - A\,x - B = 0.$$

The roots r and s may be real or complex. Complex roots are of no interest to us. We also assume $r \neq s$ and α & β constants to be determined later.

Replacing this solution-to-be into the left and the right hand side of the recurrence relation above, we have:

$$\alpha\, r^{k+2} + \beta\, s^{k+2} = \alpha\, A\, r^{k+1} + \beta\, A\, s^{k+1} + \alpha\, B\, r^k + \beta\, B\, s^k$$

or

$$\alpha\, [\, r^2 - A\, r - B\,]\ r^k = -\beta\, [\, s^2 - A\, s - B\,]\ s^k.$$

To have this expression an identity for any value of A, B, α, β, we need to choose r & s as the roots of the *characteristic equation* above as, in fact, they already are. Thus,

$$x_k = \alpha\, r^k + \beta\, s^k$$

a sum of exponentials is the closed solution where the constants α & β are still free to be selected. We choose α & β such that the recurrence relation and the closed solution have the same initial values x_0 & x_1 .

So, from the closed solution,

$$k = 0 \quad \rightarrow \quad \alpha\, r^0 + \beta\, s^0 = x_0 , \qquad k = 1 \quad \rightarrow \quad \alpha\, r + \beta\, s = x_1 ,$$

we have

$$\alpha = [x_1 - s\, x_0] \,/\, [r - s]$$
$$\beta = [x_1 - r\, x_0] \,/\, [s - r]$$

which is fine since we assumed $r \neq s$. Replacing, we get,

$$x_k = \frac{x_1 - s\, x_0}{r - s}\; r^k - \frac{x_1 - r\, x_0}{r - s}\; s^k$$

as the general solution for $r \neq s$, x_0 & x_1 given.

Let us now look into the special case $r = s$.

To find the solution for $r = s$ we are going to follow *Case* 3 and *Case* 4 of Chapter "Simple DDS with exponential growth" with recurrence relations similar and one where the exponential solutions change in analytical form when two of their parameters are different or equal in value.

Following the general solution of *Case* 4 we shall try now as solution of the special case a modified sum of exponentials

$$x_k = \alpha\, r^k + \beta\, k\, r^k$$

where r is the double root of the *characteristic equation* $x^2 - A\, x - B = 0$ of the recurrence relation

$$x_{k+2} = A\, x_{k+1} + B\, x_k .$$

The roots are given by

$$r_{1,2} = \tfrac{1}{2}\, [\, A \pm [A^2 + 4\, B]^{\frac{1}{2}}\,]$$

and a double one needs $A^2 + 4\, B = 0$ giving $r_{1,2} = \tfrac{1}{2}\, A$.

Replacing this solution-to-be into the left and the right hand side of the recurrence relation above, we have:

$$\alpha\, r^{k+2} + \beta\,[k+1]\, r^{k+2} = \alpha\, A\, r^{k+1} + \beta\, A\,[k+1]\, r^{k+1} + \alpha\, B\, r^{k} + \beta\, B\, k\, r^{k}$$

or

$$\alpha\,[\,r^2 - A\, r - B\,]\, r^{k} = -\beta\,[\,k\,[\,r^2 - A\, r - B\,] + r\,[\,2\, r - A\,]\,]\, r^{k}.$$

To have this expression an identity for any value of α & β, we need to choose r as the double root of the *characteristic equation* **and** of equation $2\, r - A = 0$ and $r = \tfrac{1}{2}\, A$ does it. Thus, for $r = \tfrac{1}{2}\, A$ with $A^2 + 4\, B = 0$,

$$x_k = \alpha\, r^{k} + \beta\, k\, r^{k}$$

is a closed solution where the constants α & β are still free to be selected. Again, to have the closed solution and the recurrence relation with the same initial values, we have from the closed solution

$$\alpha + 0 = x_0 \qquad \rightarrow \qquad \alpha = x_0$$
$$[\alpha + \beta]\, r = x_1 \qquad \rightarrow \qquad \beta = [\,x_1 - r\, x_0\,]\,/\,r\,.$$

Replacing, we obtain, for $r = s$, with x_0 & x_1 given, the closed solution

$$\mathbf{x}_k = x_0\, \mathbf{r}^{k} + [\,x_1 - r\, x_0\,]\, \mathbf{k}\, \mathbf{r}^{k-1}.$$

Non-homogeneous relations

Next, we consider our first non-homogeneous 2nd order relation

$$x_{k+2} = A\, x_{k+1} + B\, x_k + C_1$$

with A, B, C_1 constant, x_0 & x_1 ($x_1 = a_1\, x_0 + b_1\, y_0 + c_1$) given and $A + B \neq 1$.

We know that the general solution for the homogeneous case is an algebraic sum of two exponentials r^{k} & s^{k}, $r \neq s$. So, we may expect the solution for the non-homogeneous one to be of the form

$$x_k = \alpha\, r^{k} + \beta\, s^{k} + \gamma$$

where the r & s are still the roots of the *characteristic equation* and α, β, γ constants to be determined later. Again, α and β are likely to be a function of x_0 and x_1, but not necessarily γ.

Replacing the solution-to-be into the left and the right hand side of the recurrence relation above, we have,

$$\alpha r^{k+2} + \beta s^{k+2} + \gamma = \alpha A r^{k+1} + \beta A s^{k+1} + A\gamma + \alpha B r^k + \beta B s^k + B\gamma + C_1$$

or

$$\alpha [r^2 - A r - B] r^k + \gamma [1 - A - B] = -\beta [s^2 - A s - B] s^k + C_1 .$$

To have this expression an identity for any value of A, B, C, α, β we need r & s as the roots of the *characteristic equation* (which they are) and

$$\gamma [1 - A - B] = C_1 \qquad \rightarrow \qquad \gamma = C_1 / [1 - A - B]$$

provided that $A + B \neq 1$ as previously stated. Thus,

$$x_k = \alpha r^k + \beta s^k + C_1 / [1 - A - B]$$

an algebraic sum of exponentials plus a constant is a closed solution where α & β are still free to be selected. Again, we choose α & β from the closed solution such that for

$$k = 0 \qquad \rightarrow \qquad \alpha r^0 + \beta s^0 + \gamma = x_0$$
$$k = 1 \qquad \rightarrow \qquad \alpha r + \beta s + \gamma = x_1$$

with x_0 given and $x_1 = a_1 x_0 + b_1 y_0 + c_1$, Solving

$$\alpha = [[x_1 - \gamma] - s [x_0 - \gamma]] / [r - s]$$
$$\beta = [[x_1 - \gamma] - r [x_0 - \gamma]] / [s - r] .$$

Replacing, we have, in all its beauty, when $r \neq s$ and $A + B \neq 1$,

$$x_k = \gamma + \frac{[x_1 - \gamma] - s [x_0 - \gamma]}{r - s} r^k - \frac{[x_1 - \gamma] - r [x_0 - \gamma]}{r - s} s^k$$

as the general solution for the non-homogeneous case with x_0 & x_1 given.

Finale . . .

We are now in the position to link the two systems of recurrence relations. In fact, given the system

$$\begin{cases} x_{k+1} = a_1 x_k + b_1 y_k + c_1 \\[2em] y_{k+1} = a_2 x_k + b_2 y_k + c_2 \end{cases} \qquad \text{with } x_0 \text{ \& } y_0 \text{ given,}$$

we may replaced it by a system of 2nd order with separated variables

$$\begin{cases} x_{k+2} = A\ x_{k+1} + B\ x_k + C_1 \\[2em] y_{k+2} = A\ y_{k+1} + B\ y_k + C_2 \end{cases}$$

with

$$\begin{aligned} A &= a_1 + b_2 & C_1 &= [1 - b_2]\ c_1 + b_1 c_2 \\ B &= a_2 b_1 - a_1 b_2 & C_2 &= [1 - a_1]\ c_2 + a_2 c_1 \end{aligned}$$

as found previously, which, for $r \neq s$, has the general solution

$$\begin{cases} x_k = \gamma_1 + \alpha_1 r^k + \beta_1 s^k \\[2em] y_k = \gamma_2 + \alpha_2 r^k + \beta_2 s^k \end{cases}$$

where $\alpha_1, \beta_1, \alpha_2, \beta_2$ are constants dependent on the initial values x_0 & y_0 and γ_1 & γ_2. These coefficients are given by

$$\begin{aligned} \gamma_1 &= C_1 / [1 - A - B] \\ \gamma_2 &= C_2 / [1 - A - B] \end{aligned} \qquad \text{whenever } A + B \neq 1$$

and

$$\begin{aligned} \alpha_1 &= [\ [x_1 - \gamma_1] - s\ [\ x_0 - \gamma_1]\ ]\ /\ [r - s] \\ \beta_1 &= [\ [x_1 - \gamma_1] - r\ [\ x_0 - \gamma_1]\ ]\ /\ [s - r] \end{aligned}$$

$$\begin{aligned} \alpha_2 &= [\ [y_1 - \gamma_2] - s\ [\ y_0 - \gamma_2]\ ]\ /\ [r - s] \\ \beta_2 &= [\ [y_1 - \gamma_2] - r\ [\ y_0 - \gamma_2]\ ]\ /\ [s - r], \end{aligned}$$

with

$$\begin{aligned} x_1 &= a_1 x_0 + b_1 y_0 + c_1 \\ y_1 &= a_2 x_0 + b_2 y_0 + c_2 \end{aligned} \qquad \text{and } x_0 \text{ \& } y_0 \text{ given.}$$

The r & s are the distinct roots of the equation $x^2 - A x - B = 0$.

Let us now consider a numerical example: Given the linear DDS by

$$
\begin{cases}
x_{k+1} = 2 x_k - 2 y_k + 1 \\[2em]
y_{k+1} = 1.12 x_k - y_k + 0.76 \qquad\qquad \text{with } x_0 = y_0 = 0,
\end{cases}
$$

find the general solution.

1. First, we determine the roots r & s of the characteristic equation.
  From

$$A = a_1 + b_2 \qquad\qquad → \qquad A = 1$$
$$B = a_2 b_1 - a_1 b_2 \qquad → \qquad B = -0.24,$$

we have $\qquad\qquad x^2 - x + 0.24 = 0 \qquad\qquad$ with roots

$$x_{1,2} = \tfrac{1}{2}[\,1 \pm 0.2\,] \qquad → \qquad r = 0.4 \ \ \& \ \ s = 0.6, \qquad r \neq s.$$

2. Next, we calculate
$$C_1 = [1 - b_2]\, c_1 + b_1 c_2 \quad → \qquad C_1 = 0.48$$
$$C_2 = [1 - a_1]\, c_2 + a_2 c_1 \quad → \qquad C_2 = 0.36.$$

3. Then we evaluate
$$\gamma_1 = C_1 / [1 - A - B] \qquad → \qquad \gamma_1 = 2$$
$$\gamma_2 = C_2 / [1 - A - B] \qquad → \qquad \gamma_2 = 1.5.$$

4. The general solution is

$$
\begin{cases}
x_k = 2 + \alpha_1\,[\,0.4\,]^k + \beta_1\,[\,0.6\,]^k \\[2em]
y_k = 1.5 + \alpha_2\,[\,0.4\,]^k + \beta_2\,[\,0.6\,]^k
\end{cases}
$$

with $\alpha_1, \beta_1, \alpha_2, \beta_2$ dependent on $\gamma_1 = 2$, $\gamma_2 = 1.5$, $x_0 = y_0 = 0$ and

$$x_1 = 2 x_0 - 2 y_0 + 1 \qquad\qquad → \qquad x_1 = 1$$
$$y_1 = 1.12 x_0 - y_0 + 0.76 \qquad → \qquad y_1 = 0.76.$$

From their definition (previous page), we finally have

$$\alpha_1 = -1 \qquad \beta_1 = -1$$
$$\alpha_2 = -0.8 \qquad \beta_2 = -0.7 .$$

In practice, we may not need to compute the whole general solution but only to determine the long term behaviour of x_k and y_k as $k = \infty$ i.e. x^* and y^* or whether the equilibrium point is stable or not (see bellow). If it is stable, then the values of x^* and y^* can be known from $x^* = \gamma_1$ and $y^* = \gamma_2$.

Equilibrium values and stability

For $r \neq s$, we have seen that the general solution of a system of two linear recurrence relations are algebraic sums of exponentials r^k & s^k plus a constant term γ_1 or γ_2.

To have bounded solutions in the long-term, we need $|r| < 1$ **and** $|s| < 1$ i.e. we need to solve the *characteristic equation* to find if its roots are real, distinct and both $|r| < 1$ **and** $|s| < 1$.

If so, the equilibrium values are

$$x^* = \lim_{k=\infty} x_k = \gamma_1 \quad \text{and} \quad y^* = \lim_{k=\infty} y_k = \gamma_2 .$$

Note that the equilibrium values are independent of the initial values x_0 & y_0 and stable.

Regarding the previous numerical example, we have,

$$|r = 0.4| < 1 \quad \text{and} \quad |s = 0.6| < 1 .$$

So, the solution is bounded with stable equilibrium values

$$x^* = 2 \quad \text{and} \quad y^* = 1.5 \qquad\qquad \text{for all } x_0 \text{ & } y_0 .$$

22. QUADRATIC 2-POPULATION DDS

In population models the simplest quadratic DDS with 2 populations sharing the same environment and competing for common resources without trying to destroy each other is

$$\begin{cases} x_{k+1} = x_k + \varepsilon_1 \left[1 - \gamma_1 \left[x_k + y_k \right] \right] x_k \\[3em] y_{k+1} = y_k + \varepsilon_2 \left[1 - \gamma_2 \left[x_k + y_k \right] \right] y_k , \end{cases}$$

with x_0 & y_0 given and the "growth" and "attrition" coefficients $\varepsilon_1, \varepsilon_2 > 0$, $\gamma_1, \gamma_2 > 0$ constant. Again each relation has a "saturation" factor of the form $[1 - \gamma [x_k + y_k]]$ followed by a pure "growth" one $\varepsilon_1 x_k$ or $\varepsilon_2 y_k$.

Note that the "saturation" factor reflects not only the presence of individuals of one's own population but also and, in equal proportion, of individuals of the other.

We shall start by computing its equilibrium values x^* and y^* i.e. the values of x_k and y_k that make the 2nd term of the right-hand side of the above recurrence relations exactly zero.

For $\gamma_1 = \gamma_2 = \gamma$, their number is infinite however $x^* + y^* = 1/\gamma$.

For $\gamma_1 \neq \gamma_2$, the only possible equilibrium points are:

$$\begin{cases} x^* = 0 \\[2em] y^* = 0 \end{cases} \text{and} \quad \begin{cases} x^* = 1/\gamma_1 \\[2em] y^* = 0 \end{cases} \text{and} \quad \begin{cases} x^* = 0 \\[2em] y^* = 1/\gamma_2 \end{cases} .$$

So, in the long-term if the non-zero equilibrium points are stable, one of the two populations tends to zero making the other a, de facto, single population.

Since no general or phase-plane solution is known for this DDS (as it was also the case for the single quadratic one) we study the long-term behaviour of the solution by discussing the stability of the unknown solution at these non-zero equilibrium points .

We start by constructing the Jacobian at an unspecified point

$$
\left[
\begin{array}{cc}
\partial_x f = 1 + \varepsilon_1 \left[1 - \gamma_1 \left[y + 2x \right] \right] & \partial_y f = -\varepsilon_1 \gamma_1 x \\[2em]
\partial_x g = -\varepsilon_2 \gamma_2 y & \partial_y g = 1 + \varepsilon_2 \left[1 - \gamma_2 \left[x + 2y \right] \right]
\end{array}
\right].
$$

At the equilibrium point ($x^* = 1/\gamma_1$, $y^* = 0$), the eigenvalues are the roots r_1 & r_2 of the determinant

$$
\begin{vmatrix}
\left[1 - \varepsilon_1 \right] - r & -\varepsilon_1 \\[2em]
0 & \left[1 + \varepsilon_2 \left[1 - \gamma_2 / \gamma_1 \right] \right] - r
\end{vmatrix} = 0
$$

or

$$
\left[1 - \varepsilon_1 - r \right] \left[1 + \varepsilon_2 \left[1 - \gamma_2 / \gamma_1 \right] - r \right] = 0
$$

i.e.

$$r_1 = 1 - \varepsilon_1 \qquad \rightarrow \quad |r_1| < 1 \quad \text{for} \quad 0 < \varepsilon_1 < 2 \, ,$$

$$r_2 = 1 + \varepsilon_2 \left[1 - \gamma_2 / \gamma_1 \right] \quad \rightarrow \quad |r_2| < 1 \quad \text{for} \quad 0 < \varepsilon_2 < 2 / \left[\gamma_2 / \gamma_1 - 1 \right]$$

provided that $\gamma_2 / \gamma_1 > 1$ or $\gamma_1 < \gamma_2$.

So, if $\gamma_1 < \gamma_2$ and the above constraints on the ε_1 & ε_2 are **both** fulfilled, the equilibrium point is stable and the x_k population tends to $x^* = 1/\gamma_1$ while the other goes to extinction, i.e. the 2-population DDS becomes, in the long-term a, de facto, single population DDS.

Similarly, at the non-zero equilibrium point ($x^* = 0$, $y^* = 1/\gamma_2$) ,

$$r_1 = 1 + \varepsilon_1 \left[1 - \gamma_1 / \gamma_2 \right] \quad \rightarrow \quad |r_1| < 1 \quad \text{for} \quad 0 < \varepsilon_1 < 2 / \left[\gamma_1 / \gamma_2 - 1 \right] ,$$

$$r_2 = 1 - \varepsilon_2 \qquad \rightarrow \quad |r_2| < 1 \quad \text{for} \quad 0 < \varepsilon_2 < 2$$

provided that $\gamma_1 / \gamma_2 > 1$ or $\gamma_1 > \gamma_2$ and, if stable, the x_k population becomes, in the long-term, extinct while the other tends to $y^* = 1/\gamma_2$.

92

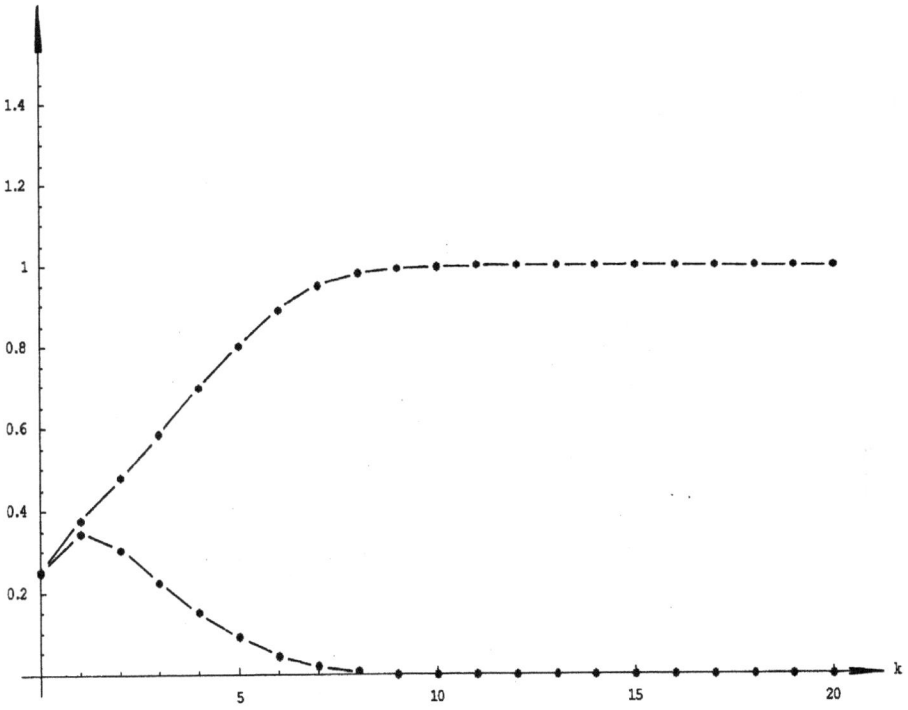

Fig. 24 Competing quadratic DDS with
$\varepsilon_1 = \gamma_1 = 1,\ \varepsilon_2 = \gamma_2 = 1.5$

More important, these DDS may not have a stable equilibrium point but, if there is one, the "growth" coefficients ε_1, ε_2 are not the parameters that determine which of the two populations survives extinction but their "attrition" coefficients γ_1, γ_2.

Fig. 24 displays the discrete solutions of a pair of competing quadratic DDS with the same $x_0 = y_0 = 0.25$. The top x_k solution with $\varepsilon_1 = \gamma_1 = 1$, the bottom y_k one with $\varepsilon_2 = \gamma_2 = 1.5$. Since $\gamma_1 < \gamma_2$, $\gamma_2 / \gamma_1 - 1 = 0.5$ we have $\varepsilon_2 < 4$ and we expect the y_k solution to go to zero smoothly even having its "growth" coefficient $\varepsilon_2 > \varepsilon_1$. This is what happens.

Instead, Fig. 25 shows the discrete solutions of a pair of competing quadratic DDS when $\gamma_1 = \gamma_2$ with the same $x_0 = y_0 = 0.25$ and $\gamma_1 = \gamma_2 = 1$. The bottom x_k solution with $\varepsilon_1 = 3$ and the top y_k one with $\varepsilon_2 = 1.5$. Both solutions come to non-zero stable equilibrium values $x^* = 0.2983$ and $y^* = 0.7017$, approximately, $x^* + y^* = 1$ as expected. A very interesting case indeed.

Note that, the x_k population would have "chaotic" behaviour on its own (in fact, it is the "chaotic" case of the quadratic DDS in Fig. 13) and how it struggles, in the presence of the second one, to stabilize at an equilibrium point $x^* < y^*$ even with $\varepsilon_1 > \varepsilon_2$.

A lot has been written about model complexity and dynamic stability (see May[10]) and, in fact, it can occur as this simple case shows.

What more can we say about the long-term behaviour of the missing algebraic solution ?

For $\gamma_1 \neq \gamma_2$ and stable equilibrium values, the solution of 2-population quadratic DDS is expected to reduce itself to the solution of a single quadratic DDS. The x_k population survives if $\gamma_1 < \gamma_2$ and the y_k population flourishes whenever $\gamma_1 > \gamma_2$.

Clearly, this may not happen if the restrictions on the growth coefficients are not fulfilled. For growth coefficients not ε_1, $\varepsilon_2 < 2$, the long-term behaviour of the populations, specially the surviving one, may be unbounded and,

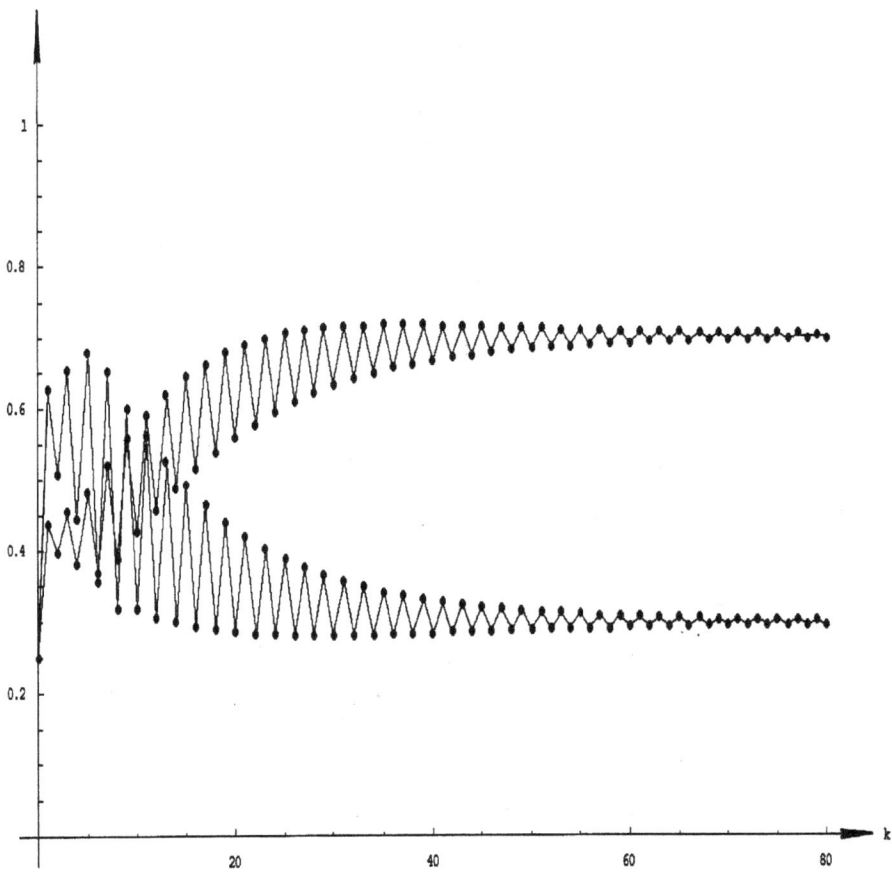

Fig. 25 Competing quadratic DDS with
$\varepsilon_1 = 3, \varepsilon_2 = 1.5, \gamma_1 = \gamma_2 = 1$

if bounded, is likely to oscillate indefinitely around the **unstable** equilibrium values $1/\gamma_1$ or $1/\gamma_2$. These oscillating solutions may be periodic with bifurcation points, as previously observed in the single quadratic DDS.

For $\gamma_1 = \gamma_2$, both populations survive like single quadratic DDS and behave as such, although we do not know beforehand their long-term equilibrium values, if any.

23. LOGISTIC 2-POPULATION CDS

The logistic CDS with 2 populations sharing the same environment and competing for common resources equivalent to the previous discrete one is:

$$\begin{cases} D_t\,y_1 = \varepsilon_1\,[1 - \gamma_1\,[\,y_1 + y_2\,]\,]\,\,y_1 \\ \\ D_t\,y_2 = \varepsilon_2\,[1 - \gamma_2\,[\,y_1 + y_2\,]\,]\,\,y_2\,, \end{cases}$$

with $y_1^0 = y_1(t_0)$ & $y_2^0 = y_2(t_0)$ given and $\varepsilon_1, \gamma_1, \varepsilon_2, \gamma_2 > 0$ constant .

For $\gamma_1 \neq \gamma_2$, the equilibrium values of this CDS are the values y_1^* and y_2^* that make $D_t\,y_1 = 0$ and $D_t\,y_2 = 0$, i.e. the values corresponding to zero "growth". The non-zero ones are

$$\begin{cases} y_1^* = 1/\gamma_1 \\ \\ y_2^* = 0 \end{cases} \quad\text{and}\quad \begin{cases} y_1^* = 0 \\ \\ y_2^* = 1/\gamma_2 \,. \end{cases}$$

So, in the long-term, the size of one of the populations nears zero and the other behaves as a single population thereafter.

The phase-plane solution

The phase-plane solution of this logistic CDS, y_1^0 & y_2^0 given, is (see ahead):

$$[\,y_1/y_1^0\,]^{\,1/[\,\varepsilon_1\gamma_1\,]} \,/\, [\,y_2/y_2^0\,]^{\,1/[\,\varepsilon_2\gamma_2\,]} = e^{\,[\,1/\gamma_1\, -\, 1/\gamma_2\,]\,[\,t\,-\,t_0\,]}.$$

Discussion of the phase-plane solution

Let us consider first $\gamma_1 \neq \gamma_2$. For $\gamma_1 > \gamma_2$, $[\,1/\gamma_1 - 1/\gamma_2\,] < 0$, the exponential tends to zero and y_1 tends also to zero as t tends to infinity since y_2 remains non zero. Alone, y_2 being a logistic population, tends to its equilibrium value $y_2^* = 1/\gamma_2$. The situation is reversed if $\gamma_1 < \gamma_2$ and, more important, the "growth" coefficients ε_1 & ε_2 are not the ones that determine which of the two populations goes to extinction but the "attrition" coefficients γ_1 & γ_2 .

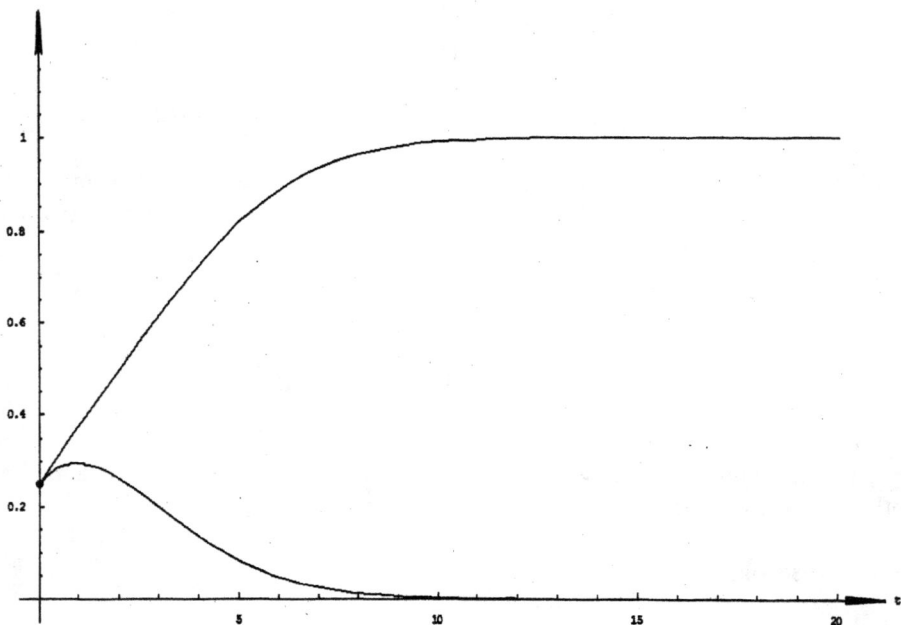

Fig. 26 Competing logistic CDS with
$\varepsilon_1 = \gamma_1 = 1, \ \varepsilon_2 = \gamma_2 = 1.5$

For $\quad \gamma_1 = \gamma_2 \qquad \rightarrow \qquad e^{\,|\,1/\gamma_1\,-\,1/\gamma_2\,|\,|\,t\,-\,t_0\,|} = 1,$

the phase-plane solution in y_1 and y_2 reduces to

$$[\, y_1 \,/\, y_1^{\,0}\,]^{\,1\,/\,|\,\varepsilon_1\,\gamma_1\,|} \;=\; [\, y_2 \,/\, y_2^{\,0}\,]^{\,1\,/\,|\,\varepsilon_2\,\gamma_2\,|}$$

and both populations survive together and forever in the same environment which is often desirable. So, it may make good sense, in practice, to adjust, whenever possible, "attrition" coefficients γ_1, γ_2 such that they are roughly equal. Otherwise, in the long-term, only one of the two populations survives.

As an example, let us generalize and consider a free market approach to an economic model. Market economists believe a free market approach produces more efficient and stable markets. However, on its own, that is not usually enough. From above we need, in addition, a Market Authority to control competition, i.e. to curtail "attrition coefficients" of some of the market players such that "attrition coefficients" are roughly the same for **all** of them. Being so, all of them can, in principle survive, obviously some better than others. Otherwise, in the long-term, only **one** of them survives and "free" market and "healthy" competition ends.

Fig. 26 exemplifies the solutions of competing logistic CDS with the same $y_1^{\,0} = y_2^{\,0} = 0.25$. The top y_1 solution with $\varepsilon_1 = \gamma_1 = 1$ and the bottom y_2 one with $\varepsilon_2 = \gamma_2 = 1.5$ as in the discrete counterpart. Since $\gamma_2 > \gamma_1$, the y_1 population grows asymptotically to the equilibrium value $y_1^* = 1$ while the y_2 population goes quickly to extinction even having its growing coefficient $\varepsilon_2 > \varepsilon_1$. This is typical of competing logistic CDS.

Instead, Fig. 27 records the solutions of competing logistic CDS when $\gamma_1 = \gamma_2$ with the same $y_1^{\,0} = y_2^{\,0} = 0.25$, $\varepsilon_1 = 3$, $\varepsilon_2 = 1.5$ and $\gamma_1 = \gamma_2 = 1$ as in the discrete counterpart. Both solutions come smoothly to non-zero stable equilibrium values $y_1^* = 0.6096$ and $y_2^* = 0.3904$, approximately, $y_1^* + y_2^* = 1$ as expected. The equilibrium values are different from the discrete counterpart ones and, more important $y_1^* > y_2^*$, not the other way around. The graphic solutions were obtained by standard numerical integration software.

Concluding, we may say that the solutions of this type of CDS and the ones of the analogue quadratic DDS are only similar if the growth coefficients are ε_1, $\varepsilon_2 < 2$. Oscillating solutions observed in 2-population quadratic DDS are out of consideration in 2-population logistic CDS.

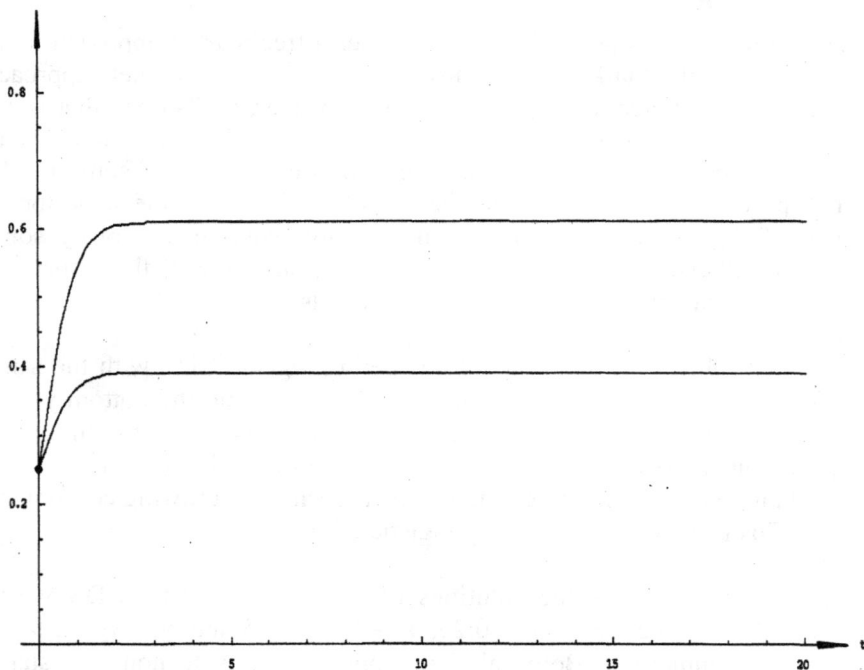

Fig. 27 Competing logistic CDS with
$\varepsilon_1 = 3$, $\varepsilon_2 = 1.5$, $\gamma_1 = \gamma_2 = 1$

24. SOLUTION OF LOGISTIC 2-POPULATION CDS

To obtain the phase-plane solution of the CDS

$$\begin{cases} D_t \, y_1 \;=\; \varepsilon_1 \, [1 \; - \gamma_1 \, [\, y_1 \, + y_2 \,]\,] \; y_1 \\[2em] D_t \, y_2 \;=\; \varepsilon_2 \, [1 \; - \gamma_2 \, [\, y_1 \, + y_2 \,]\,] \; y_2 \, , \end{cases}$$

with $y_1{}^0 = y_1(t_0)$ & $y_2{}^0 = y_2(t_0)$ given and $\varepsilon_1, \gamma_1, \varepsilon_2, \gamma_2 > 0$ constant, we start by writing it as

$$\begin{cases} [\varepsilon_1 \, \gamma_1 \, y_1]^{-1} \, D_t \, y_1 \;=\; 1/\gamma_1 - [\, y_1 \, + y_2 \,] \\[2em] [\varepsilon_2 \, \gamma_2 \, y_2]^{-1} \, D_t \, y_2 \;=\; 1/\gamma_2 - [\, y_1 \, + y_2 \,] \end{cases}$$

and

$$[\varepsilon_1 \, \gamma_1 \, y_1]^{-1} \, D_t \, y_1 \; - \; [\varepsilon_2 \, \gamma_2 \, y_2]^{-1} \, D_t \, y_2 \;=\; 1/\gamma_1 - 1/\gamma_2 \; .$$

Integrating

$$[\varepsilon_1 \, \gamma_1]^{-1} \, \ln y_1 \; - \; [\varepsilon_2 \, \gamma_2]^{-1} \, \ln y_2 \;=\; [1/\gamma_1 - 1/\gamma_2] \, t \; + K^{\#}$$

or

$$\ln y_1{}^{1/[\varepsilon_1 \gamma_1]} \; - \; \ln y_2{}^{1/[\varepsilon_2 \gamma_2]} \;=\; [1/\gamma_1 - 1/\gamma_2] \, t \; + K^{\#}$$

and

$$y_1{}^{1/[\varepsilon_1 \gamma_1]} \, / \, y_2{}^{1/[\varepsilon_2 \gamma_2]} \;=\; K \, e^{\,[1/\gamma_1 \, - 1/\gamma_2] \, t}$$

with

$$K \;=\; [\, y_1{}^0]^{\,1/[\varepsilon_1 \gamma_1]} \, / \, [\, y_2{}^0]^{\,1/[\varepsilon_2 \gamma_2]} \; e^{\,-[1/\gamma_1 \, - 1/\gamma_2] \, t_0} \quad .$$

Replacing, we have the phase-plane solution previously stated

$$[\, y_1 \, / \, y_1{}^0]^{\,1/[\varepsilon_1 \gamma_1]} \, / \, [\, y_2 \, / \, y_2{}^0]^{\,1/[\varepsilon_2 \gamma_2]} \;=\; e^{\,[1/\gamma_1 \, - 1/\gamma_2][\, t - t_0\,]} \, ,$$

$y_1{}^0$ & $y_2{}^0$ given, which is the solution for two "friendly" populations sharing the same environment .

25. LOTKA-VOLTERRA PREDATOR-PREY CDS

From a mathematical point of view it is even more interesting to consider the 2-population CDS when one of the two populations destroys the other for environmental resources.

These are the predator-prey CDS and are defined by the system of differential equations

$$\begin{cases} D_t y_1 = \varepsilon_1 [1 - \gamma_1 y_1 - \lambda_1 y_2] \, y_1 \\ \\ D_t y_2 = -\varepsilon_2 [1 - \gamma_2 y_2 - \lambda_2 y_1] \, y_2 \end{cases}$$

with $y_1^0 = y_1(t_0)$ & $y_2^0 = y_2(t_0)$ given and $\varepsilon_1, \gamma_1, \lambda_1, \varepsilon_2, \gamma_2, \lambda_2 > 0$ constant. Here the "saturation" factors $[1 - \gamma_1 y_1 - \lambda_1 y_2]$ and $[1 - \gamma_2 y_2 - \lambda_2 y_1]$ reflect the presence of individuals of both populations and in different proportions.

The population y_1 corresponds to the prey population which would grow like the original logistic CDS if population y_2 was not present. Instead, the predator population y_2 with $-\varepsilon_2 < 0$ would become extinct without the supply of prey individuals for environmental resources.

Further, in order to introduce a clear case of a predator-prey situation we assume, in addition, that the growth of the prey population y_1 is dominated by the presence of the predators, i.e. that the environment would not be taxed by the mere presence of the prey on their own .

Then, we can reasonably neglect the effect of the coefficient γ_1 by comparison with λ_1 and, similarly, disregard γ_2 by comparison with λ_2 .

So the previous system of differential equations simplifies to what is known as the Lotka-Volterra predator-prey CDS

$$\begin{cases} D_t y_1 = \varepsilon_1 [1 - \lambda_1 y_2] \, y_1 & [\, y_1 \diamond \text{prey}] \\ \\ D_t y_2 = -\varepsilon_2 [1 - \lambda_2 y_1] \, y_2 & [y_2 \diamond \text{predator}] \end{cases}$$

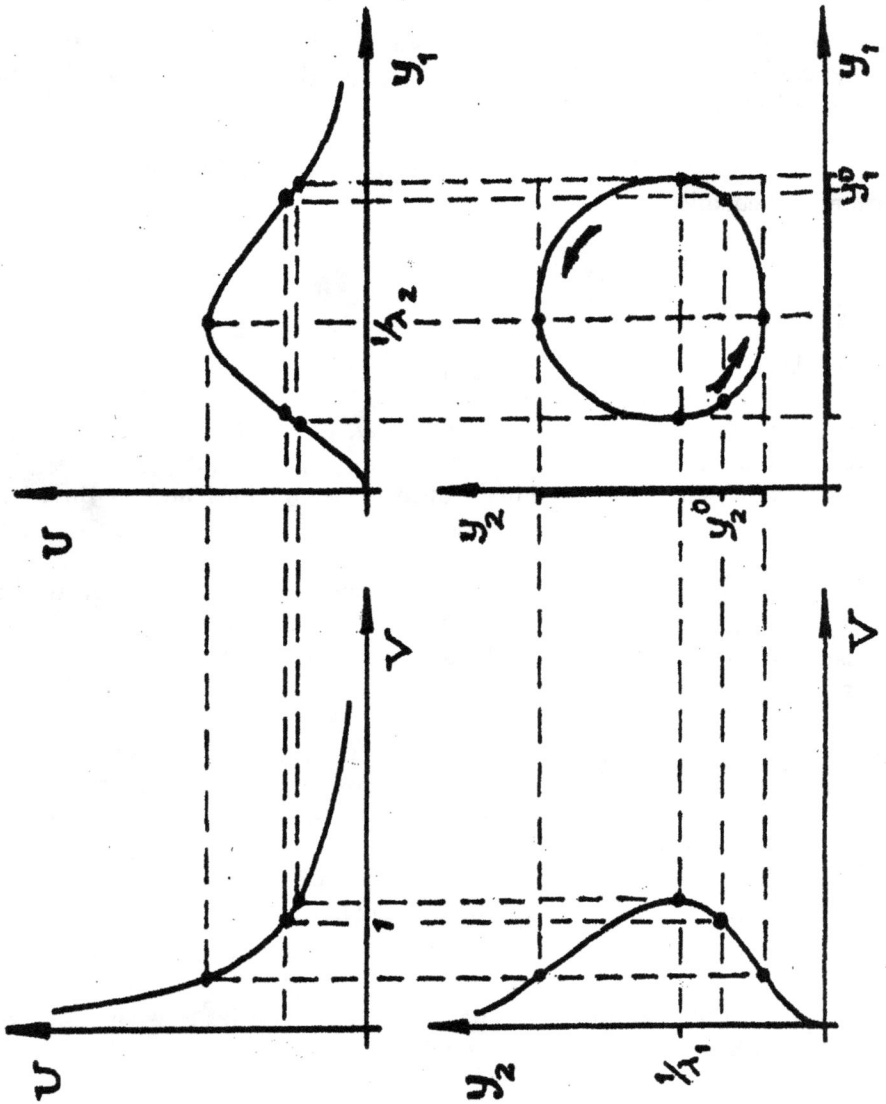

Fig. 28 Schematic Lotka-Volterra predator-prey fluctuation cycles

with $y_1^0 = y_1(t_0)$ & $y_2^0 = y_2(t_0)$ given and ε_1, λ_1, ε_2, $\lambda_2 > 0$ constant.

The phase-plane solution

The phase-plane solution is (see ahead):

$$[y_1 / y_1^0]^{1/\varepsilon_1} [y_2 / y_2^0]^{1/\varepsilon_2} = e^{[\lambda_2/\varepsilon_1][y_1 - y_1^0]} e^{[\lambda_1/\varepsilon_2][y_2 - y_2^0]}$$

which can also be written as

$$\underbrace{[[y_1 / y_1^0] e^{-\lambda_2[y_1 - y_1^0]}]^{1/\varepsilon_1}}_{U(y_1)} \underbrace{[[y_2 / y_2^0] e^{-\lambda_1[y_2 - y_2^0]}]^{1/\varepsilon_2}}_{V(y_2)} = 1.$$

This solution oscillates periodically and indefinitely .

To understand this phase-plane solution, we start by studying the shape of the factors $U(y_1)$ & $V(y_2)$ above or, to make matters easier, of U^{ε_1} & V^{ε_2}. We have,

$$D_{y_1} U^{\varepsilon_1} = [[1 - \lambda_2 y_1] / [y_1^0]] e^{-\lambda_2[y_1 - y_1^0]} .$$

which is zero at $y_1 = 1/\lambda_2$ where U^{ε_1} has a maximum.

Since at $y_1 = 0$, $U = U^{\varepsilon_1} = 0$, the function $U(y_1)$ starts growing at the origin, gets to a maximum at $y_1 = 1/\lambda_2$ and then decreases asymptotically to zero.The same can be said for $V(y_2)$. The difference being that its maximum is at $y_2 = 1/\lambda_1$.

Fig. 28 gives a visualization of $U(y_1)$, $V(y_2)$ and the corresponding *fluctuation cycle* of the phase-plane solution.

Graphic interpretation of the phase-plane solution

From Fig. 28 we guess that the solution oscillates indefinitely in a closed orbit. The closed curve of the phase-plane solution is known as the *fluctuation cycle* of the Lotka-Volterra eco-system. In more detail we have:
1.   Starting at (y_1^0, y_2^0), both populations y_1 & y_2 increase in size;
2.   Next, the predator y_2 grows at the expense of the prey y_1 and, by destroying the prey, it reaches its maximum size;

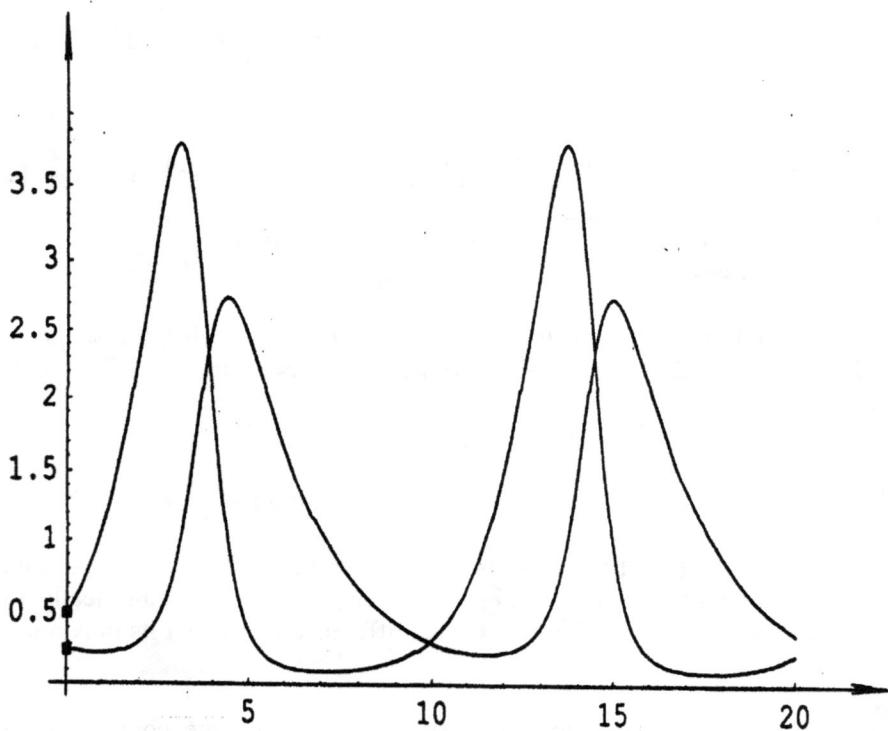

Fig. 29 Lotka-Volterra predator-prey CDS with
$\varepsilon_1 = 1$, $\varepsilon_2 = 0.5$, $\lambda_1 = \lambda_2 = 1$

3. Then by lack of prey y_1, the predator y_2 decreases in size and, in due course, reaches its minimum size;
4. With the predator y_2 at its minimum size both prey & predator grow and the closed predator-prey cycle starts again .

It can be shown that the bigger the growth coefficients the shorter the period of oscillations (e.g. Oliveira-Pinto & Conolly [12,p38]) and the larger their amplitude.

Also:
1. If the Lotka-Volterra coefficients $\varepsilon_1, \lambda_1, \varepsilon_2, \lambda_2$ change, the *fluctuation cycle* changes too and to a distinct one and the change can be dramatic (see below);
2. If the *fluctuation cycle* touches one of the axes, one of the 2 populations becomes extinct and the predator-prey CDS comes to an halt ;
3. If the predator vanishes, the prey grows exponentially to infinity !

So, these predator-prey CDS are known to be "ill-conditioned", i.e. small variations in its parameters/coefficients can produce large variations in their *fluctuation cycles* and that can be fatal.

Fig. 29 exemplifies with greater precision the periodic behaviour of the solution prey (top) predator (bottom) of a simple Lotka-Volterra CDS with $y_1^0 = 0.5$ & $y_2^0 = 0.25$ and "growth" coefficients $\varepsilon_1 = 1$, $\varepsilon_2 = 0.5$ and "attrition" coefficients $\lambda_1 = \lambda_2 = 1$, as a time function. Numerical integration techniques were used to obtain this solution.

Let us consider now a simpler "predator-prey" CDS given by the system of linear differential equations

$$\left\{ \begin{array}{l} D_t y_1 = - \varepsilon_1 \ y_2 \\ \\ \\ D_t y_2 = \ \varepsilon_2 \ y_1 \end{array} \right.$$

with $y_1^0 = y_1(t_0)$ & $y_2^0 = y_2(t_0)$ given and $\varepsilon_1, \varepsilon_2 > 0$ constant, where the "growth" of y_1 depends directly on the size of y_2 and the "growth" of y_2 on the size of y_1. For the "prey", the "growth" is negative.

The phase-plane solution is with y_1^0 & y_2^0 given,

$$\varepsilon_2 \, [\, y_1 \,]^2 + \varepsilon_1 \, [\, y_2 \,]^2 = K \qquad \text{with} \qquad K = \varepsilon_2 \, [\, y_1^0 \,]^2 + \varepsilon_1 \, [\, y_2^0 \,]^2$$

which is a set of homothetic ellipses.

In fact, re-writing the above equations as:

$$\left\{ \begin{array}{l} \varepsilon_2 \, y_1 \, D_t y_1 \;=\; - \, \varepsilon_1 \, \varepsilon_2 \; y_1 \, y_2 \\ \\ \\ \varepsilon_1 \, y_2 \, D_t y_2 \;=\; \varepsilon_1 \, \varepsilon_2 \; y_1 \, y_2 \end{array} \right.$$

and adding, we obtain

$$\varepsilon_2 \, y_1 \; D_t y_1 + \varepsilon_1 \, y_2 \; D_t y_2 = 0 \, .$$

Integrating, we get the phase-plane solution above.

So, *fluctuaction cycles* exist with amplitudes dependent on the "growth" coefficients ε_1 and ε_2. Different coefficients produce different *fluctuaction cycles* as in the Lotka-Volterra case.

Being linear, this system of differential equations can be integrated in time t to give the periodic solutions

$$\left\{ \begin{array}{l} y_1 \;=\; K_1 \; \varepsilon_1^{\,\frac{1}{2}} \; \cos \left(\, [\varepsilon_1 \, \varepsilon_2]^{\,\frac{1}{2}} \, [t - t_0] \, + K_1^{\#} \, \right) \\ \\ \\ y_2 \;=\; K_2 \; \varepsilon_2^{\,\frac{1}{2}} \; \sin \left(\, [\varepsilon_1 \, \varepsilon_2]^{\,\frac{1}{2}} \, [t - t_0] \, + K_2^{\#} \, \right) \end{array} \right.$$

where K_1, $K_1^{\#}$ & K_2, $K_2^{\#}$ are constants of integration, $t \geq t_0$. Here the solutions are simple sines / cosines with period $2\pi / [\varepsilon_1 \, \varepsilon_2]^{\,\frac{1}{2}}$. Again, the greater the growth coefficients the shorter the period of oscillations and the larger their amplitude.

Note that the solutions oscillate around zero, so y_1 and y_2 have positive and negative values. Clearly this solution can not be considered as a predator-prey solution. However, it is relevant from a mathematical point of view.

The Lotka-Volterra predator-prey CDS share the same type of periodic solutions although the solutions are positive along the whole *fluctuaction cycle*.

Historically, predator-prey models, not necessarily of the types just described, were first formulated with some success by Ross [15, p679] in 1911 to explain malaria in mankind. Ross considered the mosquito carriers of malaria as the predators and mankind as their prey.

It was however Martini [9] in 1921, who stated for the first time a set of predator-prey equations as written above to explain some predator-prey steady states. Martini was followed by Lotka [7, p88-97] in 1925 who attempted to provide a general theory for simple predator-prey situations. Lotka [6] was the first to compute, in 1923, the period of small oscillations in the case of predator-prey populations.

At the same time and, apparently, independently, Volterra [17]&[18] developed in 1927 the simple predator-prey model described above and proceeded to extend it to multi-predator-prey situations.

26. SOLUTION OF THE LOTKA-VOLTERRA CDS

To obtain the phase-plane solution of the Lotka-Volterra CDS, with y_1^0 & y_2^0 given, we start by re-writing it twice as

$$\begin{cases} [\,\varepsilon_1\, y_1\,]^{-1}\ D_t y_1\ =\ 1 - \lambda_1\, y_2 \\[2em] -[\,\varepsilon_2\, y_2\,]^{-1}\ D_t y_2\ =\ 1 - \lambda_2\, y_1 \end{cases}$$

and also as

$$\begin{cases} [\,\lambda_2\, /\, \varepsilon_1\,]\ D_t y_1\ =\ \lambda_2\ y_1\ - \lambda_1\, \lambda_2\ y_1\, y_2 \\[2em] -[\,\lambda_1\, /\, \varepsilon_2\,]\ D_t y_2\ =\ \lambda_1\ y_2\ - \lambda_1\, \lambda_2\ y_1\, y_2\ . \end{cases}$$

These systems of equations can be combined to give

$$\begin{cases} [\,\varepsilon_1\, y_1\,]^{-1}\ D_t y_1\ +\ [\,\varepsilon_2\, y_2\,]^{-1}\ D_t y_2\ =\ \lambda_2\ y_1\ - \lambda_1\, y_2 \\[1.5em] [\,\lambda_2\, /\, \varepsilon_1\,]\ D_t y_1\ +[\,\lambda_1\, /\, \varepsilon_2\,]\ D_t y_2\ =\ \lambda_2\ y_1\ - \lambda_1\, y_2\ . \end{cases}$$

Equating
$$D_t\,(\,[\,\varepsilon_1\,]^{-1}\ \ln y_1\ +[\,\varepsilon_2\,]^{-1}\ \ln y_2\,) =\ D_t\,(\,[\,\lambda_2\, /\, \varepsilon_1\,]\ y_1\ +[\,\lambda_1\, /\, \varepsilon_2\,]\ y_2\,)\ .$$

Integrating
$$[\,\varepsilon_1\,]^{-1}\ \ln y_1\ +[\,\varepsilon_2\,]^{-1}\ \ln y_2\ =\ [\,\lambda_2\, /\, \varepsilon_1\,]\ y_1\ +[\,\lambda_1\, /\, \varepsilon_2\,]\ y_2\ +\ K^{\#}$$

or
$$[\,y_1\,]^{1/\varepsilon_1}\ [\,y_2\,]^{1/\varepsilon_2} = K\ e^{[\,\lambda_2/\varepsilon_1\,]\,y_1}\ e^{[\,\lambda_1/\varepsilon_2\,]\,y_2}$$

with
$$K\ =\ [\,y_1^0\,]^{1/\varepsilon_1}\ [\,y_2^0\,]^{1/\varepsilon_2}\ e^{-[\,\lambda_2/\varepsilon_1\,]\,y_1^0}\ e^{-[\,\lambda_1/\varepsilon_2\,]\,y_2^0}\ .$$

Replacing

$$[\,y_1/\,y_1^0\,]^{1/\varepsilon_1}\ [\,y_2/\,y_2^0\,]^{1/\varepsilon_2}\ =\ e^{[\,\lambda_2/\varepsilon_1\,]\,[\,y_1 - y_1^0\,]}\ e^{[\,\lambda_1/\varepsilon_2\,]\,[\,y_2 - y_2^0\,]},$$

we have the phase-plane solution of the Lotka-Volterra predator-prey CDS which oscillates periodically and indefinitely.

27. PREDATOR-PREY DDS

We shall consider now the predator-prey DDS, given by

$$
\begin{cases}
x_{k+1} = x_k + \varepsilon_1 [1 - \lambda_1 \, y_k] \, x_k & [y_1 \diamond \text{prey}] \\
\\
y_{k+1} = y_k - \varepsilon_2 [1 - \lambda_2 \, x_k] \, y_k & [y_2 \diamond \text{predator}]
\end{cases}
$$

as the analogue of the previous Lotka-Volterra continuous one with x_0 & y_0 given and $\varepsilon_1, \lambda_1, \varepsilon_2, \lambda_2 > 0$ constant.

The non-zero equilibrium points of this dynamical system are :

$$
\begin{cases}
x^* = 1/\lambda_2 \\
\\
y^* = 0
\end{cases}
\quad \text{and} \quad
\begin{cases}
x^* = 0 \\
\\
y^* = 1/\lambda_1
\end{cases}
\quad \text{and} \quad
\begin{cases}
x^* = 1/\lambda_2 \\
\\
y^* = 1/\lambda_1
\end{cases}
$$

which we suspect to be unstable from the essentially periodic behaviour of the continuous counterpart. Let us confirm that.

We start from the Jacobian at an unspecified point

$$
\begin{bmatrix}
\partial_x f = 1 + \varepsilon_1 [1 - \lambda_1 y] & \partial_y f = - \varepsilon_1 \lambda_1 x \\
\\
\partial_x g = \varepsilon_2 \lambda_2 y & \partial_y g = 1 - \varepsilon_2 [1 - \lambda_2 x]
\end{bmatrix} .
$$

At the equilibrium point $(x^* = 1/\lambda_2, \ y^* = 0)$, we have :

$$
\begin{vmatrix}
[1 + \varepsilon_1] - r & - \varepsilon_1 \lambda_1 / \lambda_2 \\
\\
0 & 1 - r
\end{vmatrix} = 0
$$

and the eigenvalues are :

$$
r_1 = 1 \qquad \text{and} \qquad r_2 = 1 + \varepsilon_1 > 1 .
$$

The equilibrium point is not stable and, for similar reasons, the equilibrium point at $(x^* = 0, y^* = 1/\lambda_1)$, is not stable too.

Finally, for $(x^* = 1/\lambda_2, y^* = 1/\lambda_1)$, we need the eigenvalues of

$$\begin{vmatrix} 1-r & -\varepsilon_1 \lambda_1 / \lambda_2 \\ \\ \varepsilon_2 \lambda_2 / \lambda_1 & 1-r \end{vmatrix} = 0$$

or

$$[1-r]^2 + \varepsilon_1 \varepsilon_2 = 0$$

i.e.

$$r_{1,2} = 1 \pm i\, \varepsilon_1^{1/2} \varepsilon_2^{1/2}, \qquad\qquad (i^2 = -1)$$

and the eigenvalues r_1 & r_2 are complex with $|r_1| > 1$ or $|r_2| > 1$ and this equilibrium point is unstable too. The solutions may be periodic or not, but their intrinsic instability tends to bring them to cross the x_k and y_k axes and, once this happens, the predator-prey DDS comes to an halt.

We shall now consider. as we did for the Lotka-Volterra CDS , a simpler linear "predator-prey" DDS where the "growth" of both"prey"and"predator" is directly proportional to the size of the other population, i.e.

$$\begin{cases} x_{k+1} = x_k - \varepsilon_1 y_k \\ \\ y_{k+1} = y_k + \varepsilon_2 x_k \end{cases} \qquad \text{or} \qquad \begin{cases} x_{k+1} = x_k - \varepsilon_1 y_k \\ \\ y_{k+1} = \varepsilon_2 x_k + y_k \end{cases}$$

to bring it into a form previously considered for linear coupled DDS. Note that the value of x_{k+1} initialy decreases and the y_{k+1} increases and x_{k+1} have to become negative for the situation to be reversed.

Being linear we know the general solution to be ($r \neq s$)

$$\begin{cases} x_k = \alpha_1 r^k + \beta_1 s^k \\ \\ y_k = \alpha_2 r^k + \beta_2 s^k \end{cases}$$

where α_1, β_1, α_2, β_2 are constants (real) and r & s the roots of the characteristic equation, in fact, they are the roots r_1 & r_2 of the determinant of its Jacobian which can be real or compkex.

Solving the characteristic equation

$$x^2 - 2x + \varepsilon_1\varepsilon_2 + 1 = 0$$

we find the roots $r_1 = r$ and $r_2 = s$ to be

$$r_{1,2} = 1 \pm i\,\varepsilon_1^{1/2}\,\varepsilon_2^{1/2} \qquad\qquad (i^2 = -1),$$

which are complex, with $|r_1| > 1$ or $|r_2| > 1$ and the equilibrium point is unstable with the solution moving away to infinity since there are no non-linear terms in the relations to prevent the solution from doing so (x_0 & $y_0 \neq 0$).

So, we are in a curious situation of knowing the form and parameters of the general solution but are unable to used it in the real world!

Again, because x_k and y_k oscillate between positive and negative values, this DDS can not be considered as a predator-prey type DDS but still important in a discrete dynamical context.

28. HYBRID PREDATOR-PREY CDS

We now consider an alternative predator-prey CDS to the Lotka-Volterra one, defined by the system of differential equations:

$$
\begin{cases}
D_t y_1 = \varepsilon_1 \, [1 - \gamma_1 y_1 - \lambda_1 y_2] \, y_1 & [\, y_1 \diamond \text{prey}] \\[2em]
D_t y_2 = - \varepsilon_2 \, [1 - \lambda_2 y_1] \, y_2 & [y_2 \diamond \text{predator}]
\end{cases}
$$

with $y_1{}^0 = y_1(t_0)$ & $y_2{}^0 = y_2(t_0)$ given and $\varepsilon_1, \varepsilon_2, \gamma_1, \lambda_1, \lambda_2 > 0$ constant. The population y_1 corresponds to the prey population which would grow in a logistic way if the predator population y_2 is not present or goes to extinction.

Note that this system of differential equations is asymmetric, not symmetric like its predecessors. The 1st equation is of the logistic CDS type, while the 2nd one of pure predator-prey nature. So, this CDS is a cross between the logistic CDS and the Lotka-Volterra predator-prey CDS previously studied.

We may also say that the Lotka-Volterra CDS is a limiting case of this more general predator-prey CDS for $\gamma_1 = 0$.

Since the general or phase-plane solution is not known, we look for the long-term behaviour of the solution at its equilibrium points.

The non-zero equilibrium points are:

$$
\begin{cases}
y_1{}^* = 1/\lambda_2 \\[1.5em]
y_2{}^* = [1 - \gamma_1/\lambda_2]/\lambda_1
\end{cases}
\qquad \text{and} \qquad
\begin{cases}
y_1{}^* = 1/\gamma_1 \\[1.5em]
y_2{}^* = 0
\end{cases}
$$

and we shall proceed to study them. Note that the latter equilibrium point is the former one for $\gamma_1 = \lambda_2$ and, so, we do not need to consider it on its own. In fact, in this latter case the predator-prey CDS becomes, in the long-term, a logistic CDS. From the equilibrium points above, note that for $y_2^* > 0$ we must have $\gamma_1 < \lambda_2$.

For the eigenvalues, the corresponding Jacobian at an unspecified point is

$$\begin{bmatrix} \varepsilon_1 [1 - 2 \gamma_1 y_1 - \lambda_1 y_2] & - \varepsilon_1 \lambda_1 y_1 \\ \\ \varepsilon_2 \lambda_2 y_2 & - \varepsilon_2 [1 - \lambda_2 y_1] \end{bmatrix} .$$

At the first non-zero equilibrium point, $\gamma_1 \neq \lambda_2$ we obtain

$$\begin{vmatrix} - [\varepsilon_1 \gamma_1 / \lambda_2] - r & - \varepsilon_1 \lambda_1 / \lambda_2 \\ \\ \varepsilon_2 [\lambda_2 - \gamma_1] / \lambda_1 & - r \end{vmatrix} = 0 .$$

The eigenvalues are

$$r_{1,2} = - \tfrac{1}{2} \varepsilon_1 \gamma_1 / \lambda_2 \pm \tfrac{1}{2} \varepsilon_1^{\frac{1}{2}} [\varepsilon_1 \gamma_1^2 + 4 \varepsilon_2 \lambda_2 \gamma_1 - 4 \varepsilon_2 \lambda_2^2]^{\frac{1}{2}} / \lambda_2$$

and, for the values of $\gamma_1 > 0$ that make these eigenvalues complex with negative real parts we need

$$\varepsilon_1 \gamma_1^2 + 4 \varepsilon_2 \lambda_2 \gamma_1 - 4 \varepsilon_2 \lambda_2^2 < 0$$

or, solving the inequality ,

$$\gamma_1 < 2 [\varepsilon_2 / \varepsilon_1] [[1 + \varepsilon_1 / \varepsilon_2]^{\frac{1}{2}} - 1] \lambda_2 .$$

We know that in continuous 2-D spaces, equilibrium points with complex eigenvalues and negative real parts generate damped oscillations towards them (e.g. Rescigno & Richardson [14, p381]). So they are stable.

Regarding the 2nd non-zero equilibrium point, a particular case of the 1st one for $\gamma_1 = \lambda_2$, the square root above for r_1 & r_2 simplifies to $\varepsilon_1^{\frac{1}{2}} \gamma_1$ and the solution no longer oscillates.

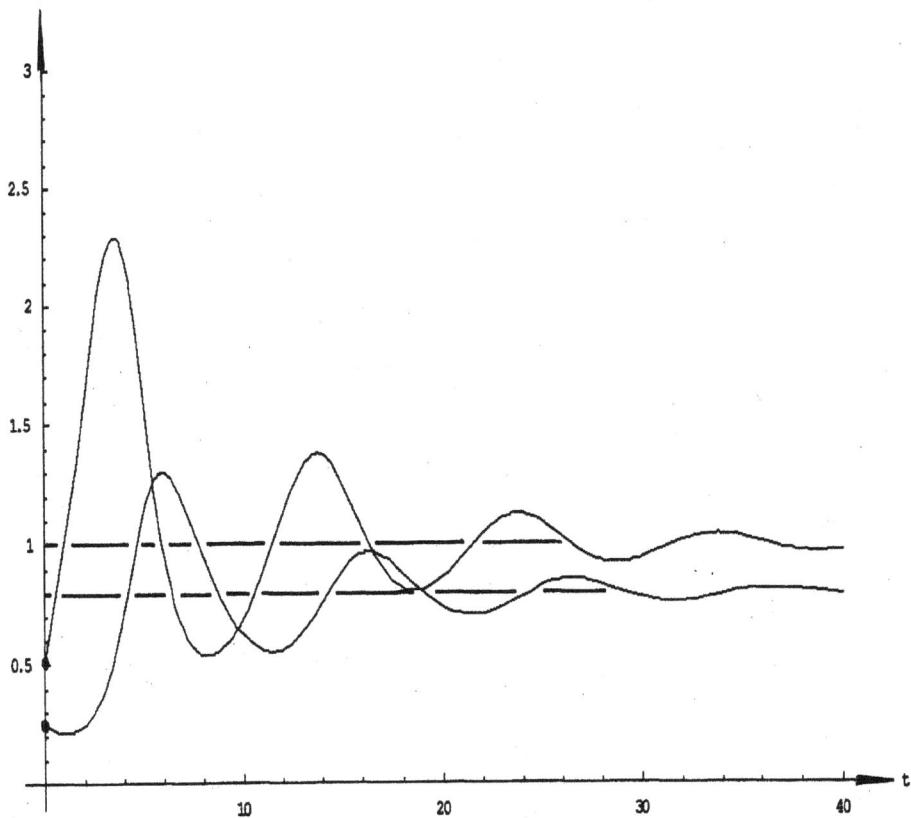

Fig. 30 Hybrid predator-prey CDS with
$\varepsilon_1 = 1,\ \varepsilon_2 = 0.5,\ \lambda_1 = \lambda_2 = 1,\ \gamma_1 = \mathbf{0.2}$

Fig. 30 shows the oscillating solution with prey (top) predator (bottom) of this type of predator-prey CDS as a time function with $y_1^0 = 0.5$ & $y_2^0 = 0.25$ and $\varepsilon_1 = 1$, $\varepsilon_2 = 0.5$, $\lambda_1 = \lambda_2 = 1$ (as in Fig. 29 of the previous Lotka-Volterra CDS example) and $\gamma_1 = \mathbf{0.2}$. The solution displays damped oscillations towards the two equilibrium points $y_1^* = 1$ and $y_2^* = 0.8$ as expected since their eigenvalues are complex, $\gamma_1 < 0.732$ (0.732 computed from inequality above) and have negative real parts.

Fig. 31 presents a predator-prey CDS with the same parameters as the previous example, prey (top) predator (bottom) except that here we took $\gamma_1 = \mathbf{0.8}$ instead of $\gamma_1 = 0.2$. Now, since $\gamma_1 > 0.732$, the solution is no longer oscillatory and tends asymptotically to the stable equilibrium values $y_1^* = 1$ and $y_2^* = 0.2$ as expected. Numerical integration techniques were used to obtain these graphic solutions.

So, about the long-term behaviour for this type of hybrid predator-prey CDS, we can say that:

1. For $\gamma_1 < \lambda_2$, the solutions tend to two equilibrium values either through damped oscillations if the eigenvalues at the equilibrium point are complex or, without damped oscillations if real.The solution can no longer oscillate indefinitely in closed orbits like in the Lotka-Volterra CDS. Clearly, this does not happen if the prey y_1 population vanishes unexpectedly.
    Note that for $\gamma_1 = 0$, this hybrid predator-prey CDS becomes the ever oscillating Lotka-Volterra one. So, for increasing values of γ_1 from 0, the oscillatory solutions decrease in magnitude until they die altogether.
2. For $\gamma_1 = \lambda_2$, the predator y_2 population becomes extinct and the prey y_1 population behaves, in the long-term, like a smooth single logistic CDS.

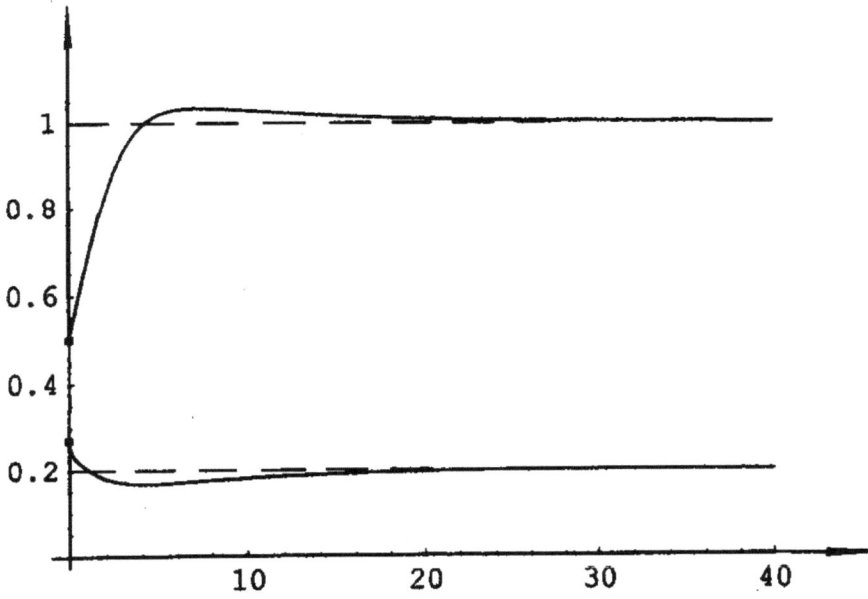

Fig.31 Hybrid predator-prey CDS with $\varepsilon_1 = 1$, $\varepsilon_2 = 0.5$, $\lambda_1 = \lambda_2 = 1$, $\boldsymbol{\gamma_1 = 0.8}$

29. HYBRID PREDATOR-PREY DDS

Now, we study the discrete analogue of the CDS just considered, i.e. the DDS given by:

$$\begin{cases} x_{k+1} = x_k + \varepsilon_1 [1 - \gamma_1 x_k - \lambda_1 y_k] x_k & [y_1 \diamondsuit \text{prey}] \\ \\ y_{k+1} = y_k - \varepsilon_2 [1 - \lambda_2 x_k] y_k & [y_2 \diamondsuit \text{predator}] \end{cases}$$

with x_0 & y_0 given and $\varepsilon_1, \gamma_1, \lambda_1, \varepsilon_2, \lambda_2 > 0$ constant.

The x_k population corresponds to the prey population which would grow in a logistic way if the predator y_k population was not present.

This system of recurrence relations is asymmetric like its predecessor with the 1st recurrence relation of the quadratic DDS type and the 2nd one of the discrete predator-prey variety.

The equilibrium points of this DDS are the ones of the continuous analogue and the non-zero ones are:

$$\begin{cases} x^* = 1/\lambda_2 \\ \\ y^* = [1 - \gamma_1/\lambda_2]/\lambda_1 \end{cases} \quad \text{and} \quad \begin{cases} x^* = 1/\gamma_1 \\ \\ y^* = 0 \end{cases} .$$

Regarding stability, we proceed to consider the Jacobian at an unspecified point

$$\begin{bmatrix} 1 + \varepsilon_1 [1 - 2\gamma_1 x - \lambda_1 y] & -\varepsilon_1 \lambda_1 x \\ \\ \varepsilon_2 \lambda_2 y & 1 - \varepsilon_2 [1 - \lambda_2 x] \end{bmatrix} .$$

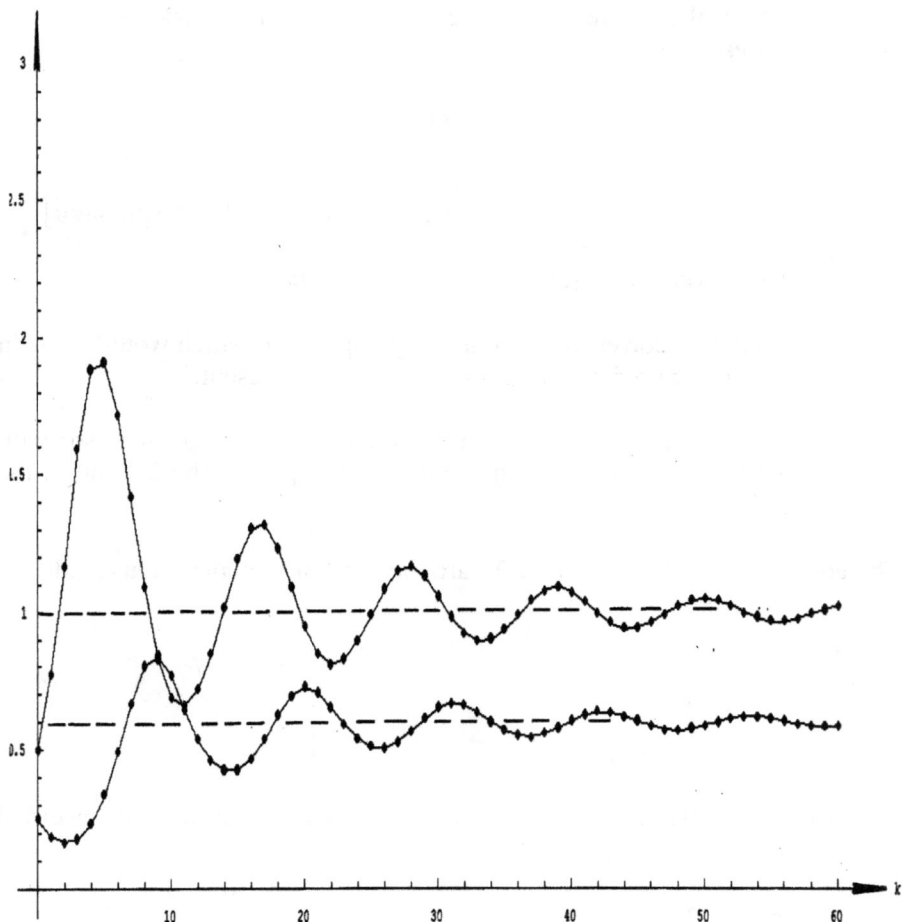

Fig. 32 Hybrid predator-prey DDS with
$\varepsilon_1 = 1$, $\varepsilon_2 = 0.5$, $\lambda_1 = \lambda_2 = 1$, $\gamma = \mathbf{0.4}$

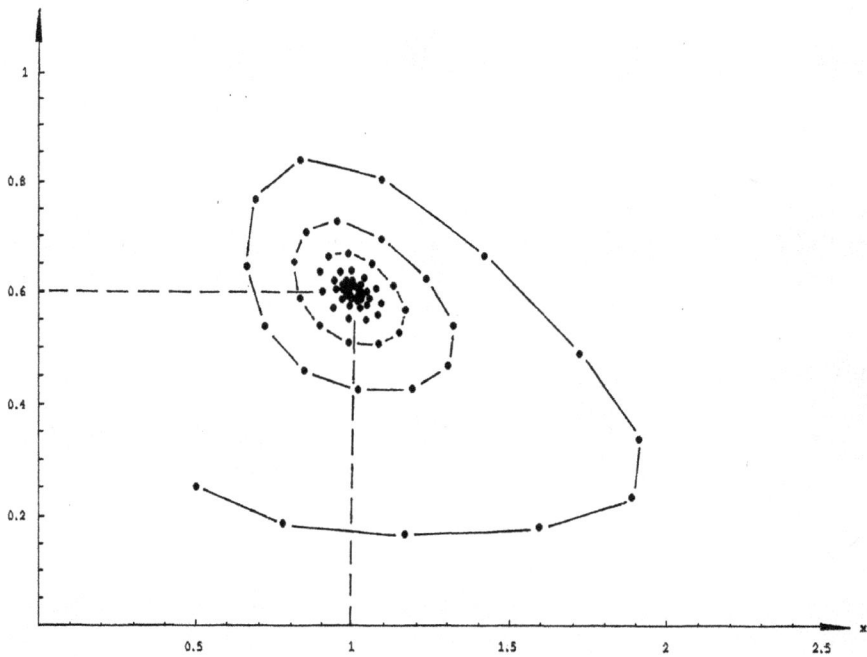

Fig. 33 Phase-plane solution of hybrid predator-prey DDS with
$\varepsilon_1 = 1, \ \varepsilon_2 = 0.5, \ \lambda_1 = \lambda_2 = 1, \ \gamma_1 = 0.4$

At the first non-zero equilibrium point, $\gamma_1 \neq \lambda_2$, we have

$$\begin{vmatrix} [1 - \varepsilon_1 \gamma_1 / \lambda_2] - r & -\varepsilon_1 \lambda_1 / \lambda_2 \\ \\ \varepsilon_2 [\lambda_2 - \gamma_1] / \lambda_1 & 1 - r \end{vmatrix} = 0$$

and the eigenvalues are similar to the ones of the continuous counterpart and given by

$$r_{1,2} = 1 - \tfrac{1}{2} \varepsilon_1 \gamma_1 / \lambda_2 \pm \tfrac{1}{2} \varepsilon_1^{\frac{1}{2}} [\varepsilon_1 \gamma_1^2 + 4 \varepsilon_2 \lambda_2 \gamma_1 - 4 \varepsilon_2 \lambda_2^2]^{\frac{1}{2}} / \lambda_2.$$

This equilibrium point is stable for a wide combination of coefficients $\varepsilon_1, \gamma_1, \lambda_1, \varepsilon_2, \lambda_2 > 0$ that make $|r_1| < 1$ **and** $|r_2| < 1$.

The values of $\gamma_1 > 0$ that make these eigenvalues complex (and the solutions to oscillate), are the same of the continuous counterpart and given by

$$\gamma_1 < 2 [\varepsilon_2 / \varepsilon_1] [[1 + \varepsilon_1 / \varepsilon_2]^{\frac{1}{2}} - 1] \lambda_2.$$

Regarding the 2nd non-zero equilibrium point, its eigenvalues are the eigenvalues above for $\gamma_1 = \lambda_2$

$$r_{1,2} = 1 - \tfrac{1}{2} \varepsilon_1 \pm \tfrac{1}{2} \varepsilon_1 \quad \rightarrow \quad r_1 = 1 \quad \text{and} \quad r_2 = 1 - \varepsilon_1$$

and the equilibrium point may be stable.

Fig. 32 displays the oscillating solution with prey (top) predator (bottom) of the hybrid predator-prey DDS with $x_0 = 0.5$ & $y_0 = 0.25$, $\varepsilon_1 = 1$, $\varepsilon_2 = 0.5$ and $\lambda_1 = \lambda_2 = 1$ (as in the previous hybrid CDS example) and $\gamma_1 = \textbf{0.4}$. Since $\gamma_1 < 0.732$ (0.732 computed from inequality above) the solution shows damped oscillations towards the two equilibrium values $x^* = 1$ and $y^* = 0.6$ as expected.

Fig. 33 gives the corresponding spiral y *versus* x phase-plane solution from the initial values to the equilibrium point $(x^* = 1, y^* = 0.6)$. Because spiral phase-plane solutions are often studied in the Natural Sciences and in Economics these hybrid dynamic systems are relevant in their understanding

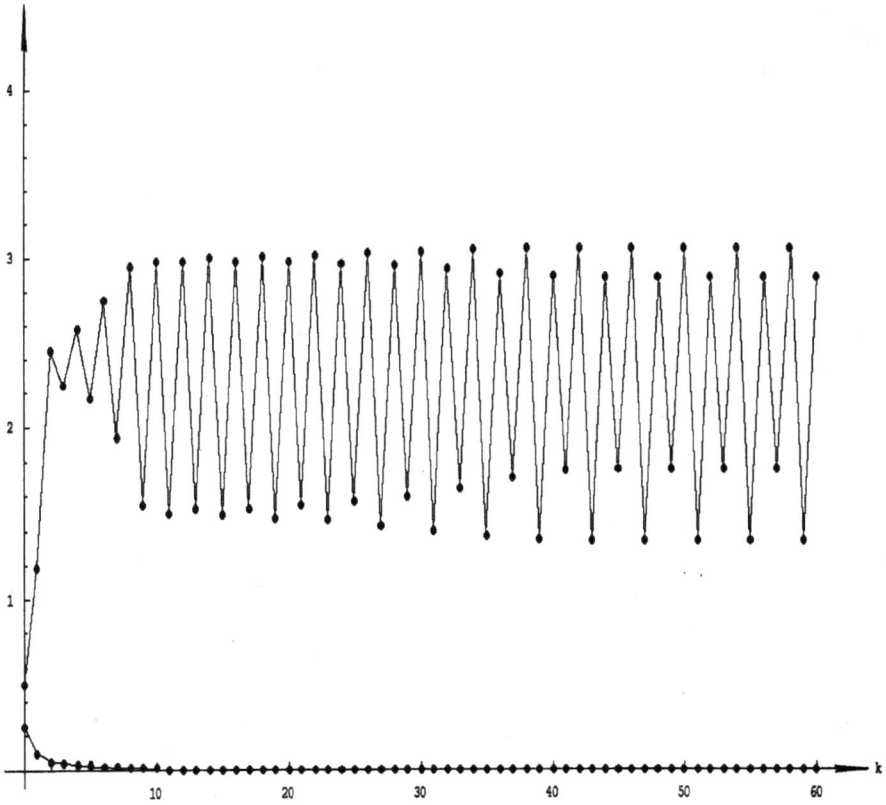

Fig. 34 Hybrid predator-prey DDS with
$\varepsilon_1 = 2.5, \varepsilon_2 = 0.5, \lambda_1 = 1, \lambda_2 = \gamma_1 = \mathbf{0.4}$

To illustrate the solution of hybrid predator-prey DDS when $\gamma_1 = \lambda_2$, we prepared Fig. 34 with $x_0 = 0.5$ & $y_0 = 0.25$, $\varepsilon_1 = 2.5$, $\varepsilon_2 = 0.5$, $\lambda_1 = 1$ and $\gamma_1 = \lambda_2 = 0.4$. The predators (bottom) smoothly approach zero, as expected, but the prey (top) tend to a periodic solution of period 4 (as in the single quadratic DDS of "growth" coefficient $\varepsilon = 2.5$ of Fig. 12).

So, the non-vanishing part of the solution does not show damped oscillations but oscillates indefinitely around the unstable equilibrium value $x^* = 2.5$. *Bifurcation points* exist which have no place in the continuous counterpart CDS.

Regarding truly periodic solutions, let us combine a naturally periodic predator-prey DDS without a stable equilibrium value (like the one of the previous example) with a "standard" predator. The coefficients are: $\varepsilon_1 = 2.5$, $\lambda_1 = 1, \gamma_1 = 0.4$ and $\varepsilon_2 = 1, \lambda_2 = 1$ with $x_0 = 0.5$ & $y_0 = 0.2$ as initial values.

Fig. 35 shows the fully periodic solution with prey (top) predator (bottom). The corresponding y *versus* x phase-plane solution, converges to a closed orbit of period 7 (an unusual period) in Fig. 36 .

In general, regarding the long-term behaviour of the solutions, we can say:

For $\gamma_1 < \lambda_2$ and stable equilibrium values, the solutions of this type of predator-prey DDS are expected to tend to a stable equilibrium point through damped oscillations if the eigenvalues at the equilibrium point are complex or, without oscillations if real. So, the oscillatory solutions decrease in magnitude until they die altogether for increasing values of γ_1 ($\gamma_1 < \lambda_2$).

Otherwise, we can only say that for unstable equilibrium values, the solutions may be unbounded and, if bounded, are likely to oscillate indefinitely periodically or not, as in the previous example where the DDS oscillates in a stable periodic manner first predicted by Kolmogorov [4] for CDS in 1936.

Needless to say that if the solution accidentally becomes zero or negative for the prey, the predator-prey DDS comes to an halt.

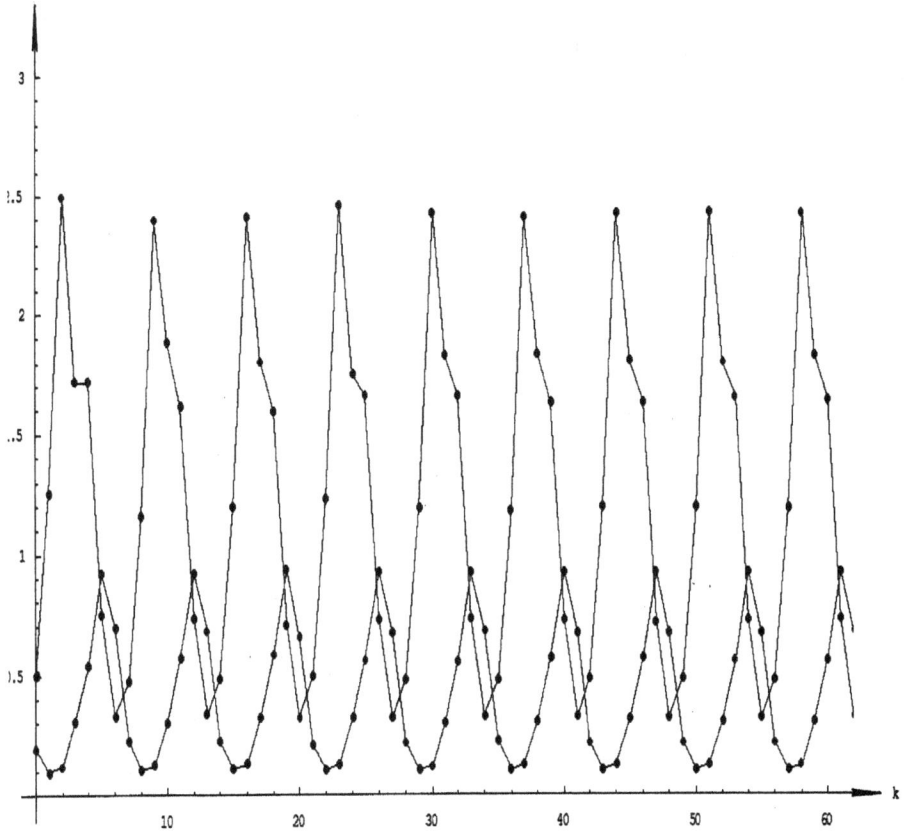

Fig. 35 Hybrid predator-prey DDS with
$\varepsilon_1 = \mathbf{2.5}, \varepsilon_2 = 1, \lambda_1 = \lambda_2 = 1, \gamma_1 = \mathbf{0.4}$

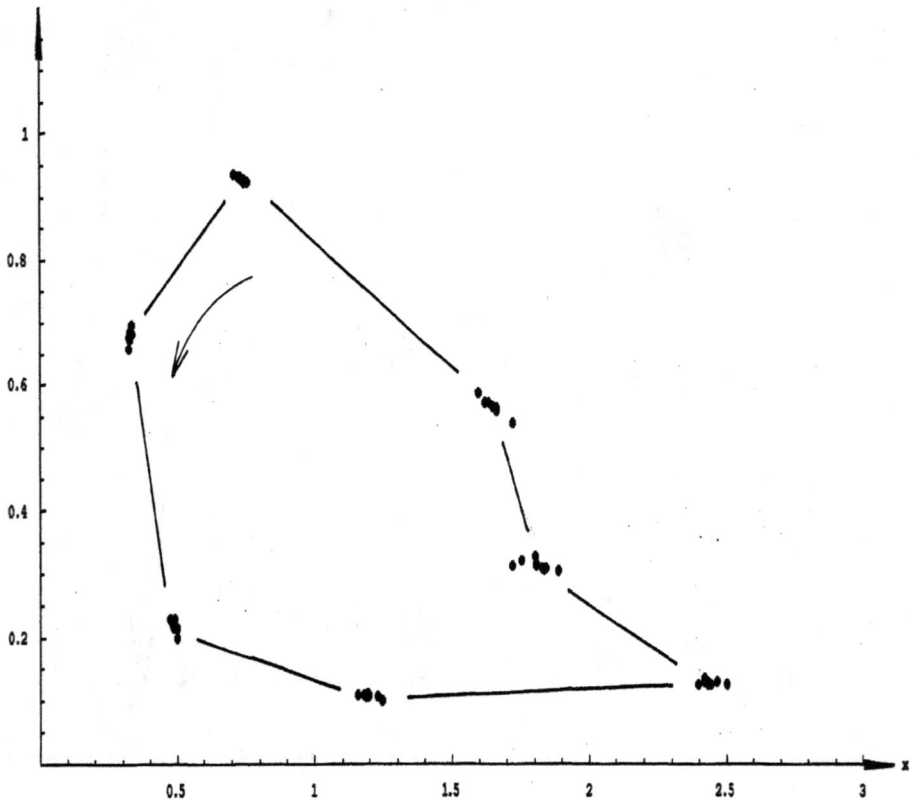

Fig. 36 Phase-plane solution of hybrid predator-prey DDS with
$\varepsilon_1 = \mathbf{2.5,}\ \ \varepsilon_2 = 1,\ \ \lambda_1 = \lambda_2 = 1,\ \ \gamma_1 = \mathbf{0.4}$

30. HUNTING / FISHING on HYBRID DDS

Finally with hybrid predator-prey DDS we want to study how the removal of some of their own individuals affects them. To us, it seems more natural to remove predators than prey and we shall do so.

Note that we prefer the expression hunting/fishing to harvesting/fishing since the word harvesting is used primarily with prey not predators.

The next question is: should we remove the DDS individuals at a constant rate, or as a proportion of the predator size?

Numerical experimentation with hybrid DDS shows that a constant hunting/fishing rate can, at times, be tricky and risky like in the 1-dimension harvesting/fishing cases previously considered. Instead, a proportional hunting/fishing effort is generally safe regarding predator/prey extinction and it leads to a more manageable analytical approach.

So, we study hybrid predator-prey DDS given by:

$$
\begin{cases}
x_{k+1} = x_k + \varepsilon_1 \left[1 - \gamma_1 x_k - \lambda_1 y_k \right] x_k & [\, y_1 \diamond \text{prey} \,] \\[2em]
y_{k+1} = y_k - \varepsilon_2 \left[1 - \lambda_2 x_k \right] y_k - \mu\, y_k & [\, y_2 \diamond \text{predator} \,]
\end{cases}
$$

with x_0 & y_0 given and $\varepsilon_1, \gamma_1, \lambda_1, \varepsilon_2, \lambda_2, \mu > 0$ constant.

The x_k population corresponds to the prey population which would grow in a quadratic way if the predator y_k population was not present. The term $\mu\, y_k$ represents the proportional hunting/fishing effort.

No general solution is known and we study again its long-term behaviour by its stability at the equilibrium points.

The non-zero equilibrium points are:

$$\begin{cases} x^* = [1+\mu/\varepsilon_2]/\lambda_2 \\ \\ y^* = [1 - \gamma_1[1+\mu/\varepsilon_2]/\lambda_2]/\lambda_1 \end{cases} \quad \text{and} \quad \begin{cases} x^* = 1/\gamma_1 \\ \\ y^* = 0 \end{cases}.$$

Note that the latter equilibrium point is the former one for $\gamma_1 = \lambda_2/[1+\mu/\varepsilon_2]$ and it is not going to be considered on its own. From above,

$$y^* > 0 \qquad \text{implies} \qquad \gamma_1 < \lambda_2/[1+\mu/\varepsilon_2]$$

and the upper bound for γ_1 depends not only on λ_2 as previously, but also on μ and ε_2 and it is smaller.

From above, for a sustainable hunting/fishing effort, we also require:

$$\mu < \varepsilon_2[\lambda_2/\gamma_1 - 1] \qquad \text{to have} \qquad y^* > 0$$

and μ is zero for $\gamma_1 = \lambda_2$.

We proceed to consider the Jacobian at an unspecified point

$$\begin{bmatrix} 1 + \varepsilon_1[1 - 2\gamma_1 x - \lambda_1 y] & -\varepsilon_1 \lambda_1 x \\ \\ \varepsilon_2 \lambda_2 y & 1 - \varepsilon_2[1 + \mu/\varepsilon_2 - \lambda_2 x] \end{bmatrix}.$$

At the first non-zero equilibrium point, $\gamma_1 \neq \lambda_2/[1+\mu/\varepsilon_2]$, we have

$$\begin{vmatrix} [1 - \varepsilon_1 \gamma_1[1+\mu/\varepsilon_2]/\lambda_2] - r & -\varepsilon_1 \lambda_1[1+\mu/\varepsilon_2]/\lambda_2 \\ \\ \varepsilon_2[\lambda_2 - \gamma_1[1+\mu/\varepsilon_2]]/\lambda_1 & 1 - r \end{vmatrix} = 0$$

and the eigenvalues are

$$r_{1,2} = 1 - \tfrac{1}{2}[\varepsilon_1/\varepsilon_2][\varepsilon_2+\mu]\gamma_1/\lambda_2 \pm \tfrac{1}{2}[\varepsilon_1/\varepsilon_2]^{1/2}[\varepsilon_2+\mu]^{1/2}$$

$$[[\varepsilon_1/\varepsilon_2][\varepsilon_2+\mu]\gamma_1^2 + 4[\varepsilon_2+\mu]\lambda_2\gamma_1 - 4\varepsilon_2\lambda_2^2]^{1/2}/\lambda_2.$$

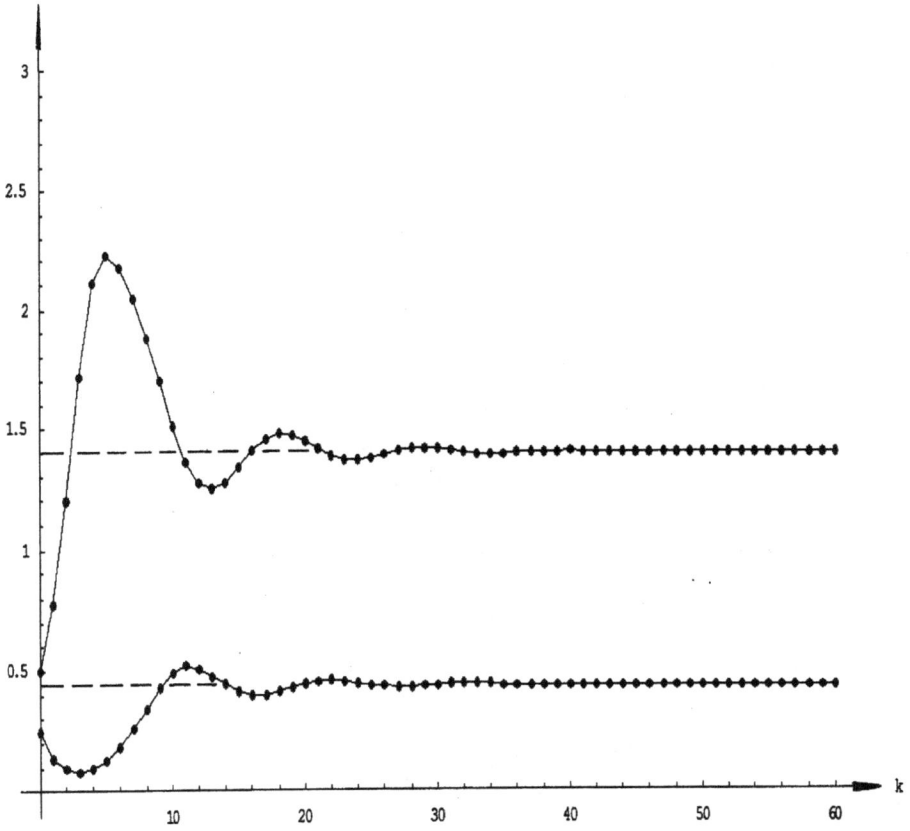

Fig. 37 Proportional hunting/fishing hybrid predator-prey DDS with
$\varepsilon_1 = 1, \ \varepsilon_2 = 0.5, \ \lambda_1 = \lambda_2 = 1, \ \boldsymbol{\gamma_1 = 0.4, \ \mu = 0.2}$,

This equilibrium point is stable for a wide combination of coefficients ε_1, γ_1, $\lambda_1, \varepsilon_2, \lambda_2, \mu > 0$ that make $|r_1| < 1$ **and** $|r_2| < 1$.

For $\mu = 0$ we obtain, as expected,

$$r_{1,2} = 1 - \tfrac{1}{2} \varepsilon_1 \gamma_1 / \lambda_2 \pm \tfrac{1}{2} \varepsilon_1^{\frac{1}{2}} [\varepsilon_1 \gamma_1^2 + 4 \varepsilon_2 \lambda_2 \gamma_1 - 4 \varepsilon_2 \lambda_2^2]^{\frac{1}{2}} / \lambda_2.$$

The values of $\gamma_1 > 0$ that make these eigenvalues complex (and the solutions to oscillate), are the ones that make

$$[\varepsilon_1 / \varepsilon_2] \gamma_1^2 + 4 \lambda_2 \gamma_1 - 4 \varepsilon_2 \lambda_2^2 / [\varepsilon_2 + \mu] < 0$$

or, solving the inequality,

$$\gamma_1 < 2 [\varepsilon_2 / \varepsilon_1] [[1 + \varepsilon_1 / [\varepsilon_2 + \mu]]^{\frac{1}{2}} - 1] \lambda_2,$$

which are similar to the ones previously found for $\mu = 0$.

Fig. 37 records the solution with prey (top) predator (bottom) of the hunting/fishing hybrid DDS with $x_0 = 0.5$ & $y_0 = 0.25$ and $\varepsilon_1 = 1, \varepsilon_2 = 0.5$, $\lambda_1 = \lambda_2 = 1$, $\gamma_1 = 0.4$ (as in Fig. 32 of the first hybrid DDS example of the previous chapter) and $\mu = 0.2$ which is fine $\mu < 0.75$ ($\mu < \varepsilon_2 [\lambda_2 / \gamma_1 - 1]$).

Since $\gamma_1 < 0.558$ (0.558 from the inequality above) the solution shows damped oscillations towards the two equilibrium values $x^* = 1.4$ and $y^* = 0.44$ as expected.

When we compare this solution with the one of the first hybrid DDS example of the previous chapter with the same parameters but $\mu = 0$, we find that here the equilibrium value for the prey x^* is larger than the one there.

In effect, for $\mu > 0$ and $\varepsilon_2 > 0$, we always have

$$x^* = [1 + \mu / \varepsilon_2] / \lambda_2 \quad \text{which is greater than} \quad x^* = 1 / \lambda_2$$

when $\mu = 0$ and the removal of predators do increase the equilibrium value of the prey which can be of value in eco-systems.

What else can we say about the long-term behaviour of the solution for this type of hunting/fishing hybrid predator-prey DDS ?

Basically the same of what was said regarding $\mu = 0$ ones but now with some modifications in the case the equilibrium values are stable. We recall:

For $\gamma_1 < \lambda_2 / [1 + \mu / \varepsilon_2]$ and stable equilibrium values, the solutions of this type of predator-prey DDS are expected to tend to a stable equilibrium point through damped oscillations if the eigenvalues at the equilibrium point are complex or, without oscillations, if real.

The difference being that by removing predators we effectively reduce their overall growth rate and that weakens the predators influence on the prey, i.e. the amplitude of the damped oscillations (if they exist) is likely to decrease and, more important, the period of the oscillations is likely to increase significantly.

So, proportional hunting/fishing on hybrid predator-prey DDS have a potential stabilizing effect on the eco-system but remember that in the limit, $\gamma_1 = \lambda_2 / [1 + \mu / \varepsilon_2]$ and $y^* = 0$, no hunting/fishing is possible in a sustainable way.

Volterra [18] studied the same hunting/fishing effect using the Lotka-Volterra CDS considered in a previous chapter and applied it to explain variations in the size of fish stocks in the Adriatic Sea during World War I.

During World War I, it was not fishing, but its absence, that produced a decrease of the prey in the Adriatic Sea.

31. GENERAL KOLMOGOROV CDS

Finally we are going to look into general CDS originally proposed by *Kolmogorov* [4] in 1936 (Oliveira-Pinto & Conolly [12, p.116-123] for an English translation) and aimed at CDS that have not only equilibrium points but also closed orbits as equilibrium curves similar to the fluctuation cycles first found in Lotka-Volterra CDS.

In an historical context, we shall follow closely young Kolmogorov speculative reasoning.

The Kolmogorov general system of differential equations is given by:

$$
\begin{cases}
D_t y_1 = F(y_1, y_2) \, y_1 & [\, y_1 \diamond \text{prey}] \\
\\
D_t y_2 = G(y, y_2) \, y_2 & [y_2 \diamond \text{predator}]
\end{cases}
$$

with $y_1^0 = y_1(t_0)$ & $y_2^0 = y_2(t_0)$ given and $F(y_1, y_2)$ & $G(y_1, y_2)$ assumed continuous functions with continuous $\partial_{y1} F$, $\partial_{y2} F$ and $\partial_{y1} G$, $\partial_{y2} G$.
The $F(y_1, y_2)$ and $G(y_1, y_2)$ describe the "growth rate" of the prey and predator populations above, respectively.

Quantitative constraints have to be imposed on these "growth rates" if the system is to reflect the predator-prey type of interaction.

To formulate these constraints more easily, we shall denote by s the direction OP from the origin O to any point $P(y_1, y_2)$ in the (y_1, y_2) phase-plane and by $d_s F$ and $d_s G$ the derivatives of $F(y_1, y_2)$ and $G(y_1, y_2)$ in this direction.

For the prey "growth rate", the constraints are:

1: $\partial_{y2} F < 0$ i.e. the prey "growth rate" decreases with increasing size of the predators;

2: $d_s F < 0$ i.e. for any given predator-prey ratio, the prey increases with decreasing rapidity either because of logistic effect on the prey and/or of an increase predator-prey encounters;

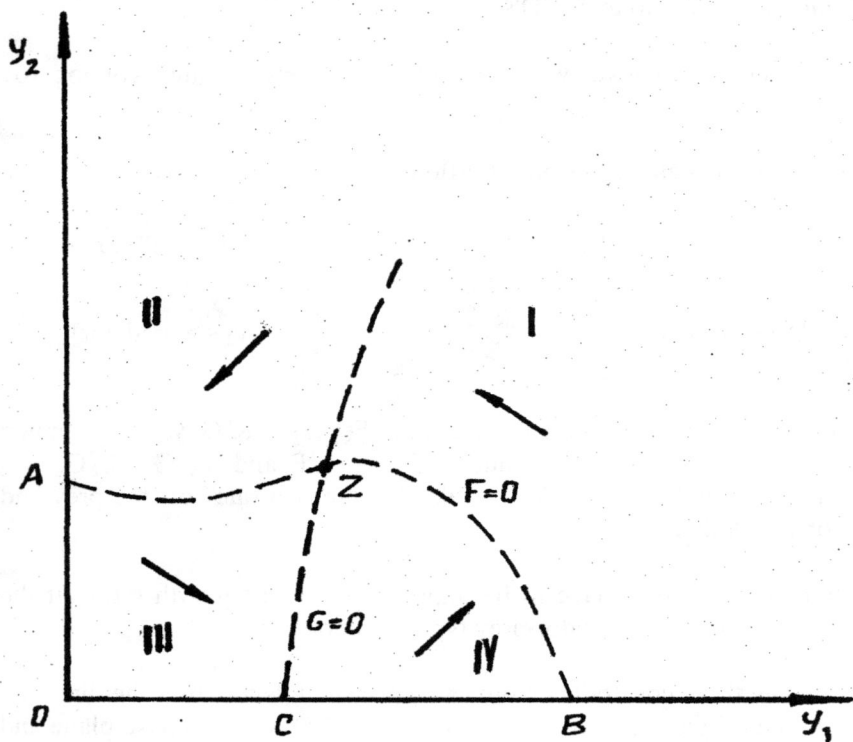

Fig. 38 Kolmogorov positive quadrant for phase-plane solutions

3: $F(0, 0) > 0$ i.e. for small sizes of both predator and prey, the prey tends to increase;

4: $F(0, A) = 0$ i.e. there is a constant $A > 0$ that for a sufficiently large predator size, the prey can no longer increase in size;

5: $F(B, 0) = 0$ i.e. there is a constant $B > 0$ such that for a sufficiently large prey size, the prey stops increasing even in the absence of predators.

Regarding the predator "growth rate", we require that:

6: $\partial_{y2} G < 0$ i.e. the predator "growth rate" decreases with their size;

7: $d_s G > 0$ i.e. for any given predator-prey ratio an increase in both populations is to the advantage of the predators;

8: $G(C, 0) = 0$ i.e. there is a critical constant $C > 0$ such that the predators can only increase if the size of the prey exceeds C with $C < B$.

Note that $G(0, D) = 0, \ D > 0$ is not mentioned otherwise non-zero predators would exist without prey.

Fig. 38 shows two possible equilibrium curves $F(y_1, y_2) = 0$ for y_1 and $G(y_1, y_2) = 0$ for y_2 ($F=0$ & $G=0$ in the Fig.) that fulfil the previous constraints with only one common equilibrium point Z where they cross. These equilibrium curves divide the quadrant in four domains **I, II, III, IV** as shown.

In domain **I**, $F(y_1, y_2) < 0$ and $G(y_1, y_2) > 0$ so, from any point $P(y_1, y_2)$ in it, the phase-plane solution decreases in y_1, i.e. goes left and increases in y_2, i.e. goes up in the quadrant. In due time, it crosses $G(y_1, y_2) = 0$ and, in domain **II**, $F(y_1, y_2) < 0$ but $G(y_1, y_2) < 0$. So, the phase-plane solution goes left and down and it crosses $F(y_1, y_2) = 0$.
In domain **III**, $F(y_1, y_2) > 0$ and in $G(y_1, y_2) < 0$ so, it goes right and down and, with luck, goes right and up in domain **IV** without crossing the predator y_2 and/or the prey y_1 axes. In other words, the phase-plane solution goes around & around the equilibrium point Z always in an anti-clockwise way.

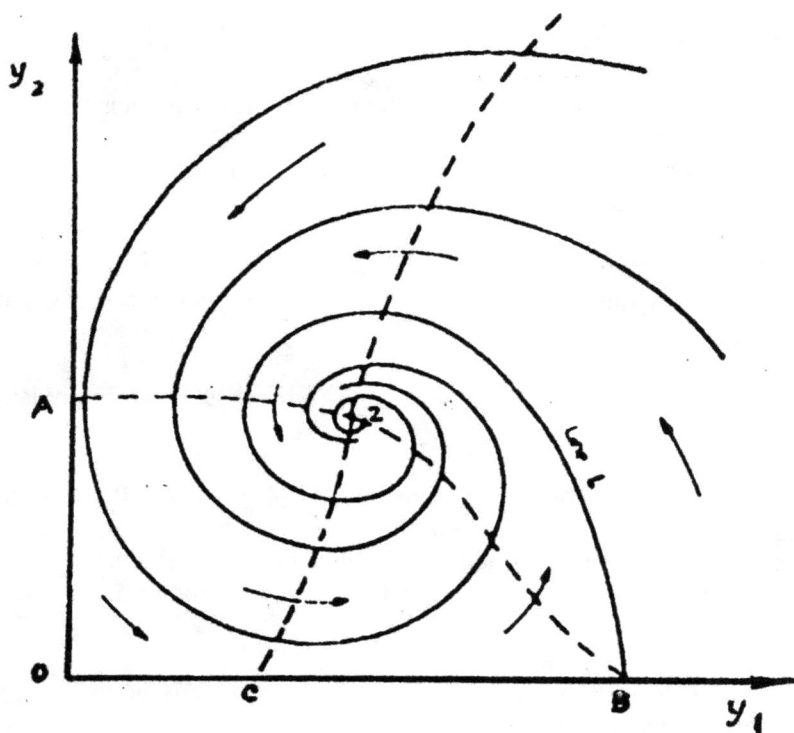

Fig.39 Schematic phase-plane solutions around stable equilibrium point Z

We now proceed to analyse, in a speculative way, the analytical behaviour of a phase-plane solution which originates at the point P(B, 0) where the predators start to grow. The phase-plane solution is denoted as curve L in Fig. 39 / Fig. 40 for case **II** and case **III** respectively. No drawing for case **I**.

Case I : The phase-plane solution approaches the stable equilibrium point Z without oscillating around it or, as Kolmogorov says, approaches Z at a definite angle.

Note this is the case of our competing 2-population CDS.

Case II : The phase-plane solution spirals infinitely many times around the stable equilibrium point Z and the solution oscillates with decreasing amplitude towards Z as exemplified in Fig. 39.

Note this is the case of some of our hybrid predator-prey CDS.

Case III: The phase-plane solution spirals infinitely many times around the **unstable** equilibrium point Z without ever getting near it, approaching instead a closed curve E that contains Z in its interior as in Fig. 40. The phase-plane solution approaches the stable equilibrium curve E from the inside or the outside with amplitude and oscillating period resembling the ones of curve E .

As far we are concerned, this is a novel case with CDS but already observed with DDS. We need algebraically more complex than logistic CDS, or logistic CDS with slowly varying coefficients (seasonal variations) for this to happen.

In nature, if the prey "growth rate" far exceeds the predator one, the tendency is for a stable closed curve to be observed (as in case **III**) instead of a stable equilibrium point solution (as in case **II**).

The Lotka-Volterra CDS is different from case **III** above because the fluctuation cycles are fixed in the sense that there are no oscillations towards them. If parameter/coefficient changes take place, a new fluctuation cycle is immediately created and the Lotka-Volterra CDS will oscillate periodically and indefinitely around it. The equivalent equilibrium point Z does not attract or repel the solution.

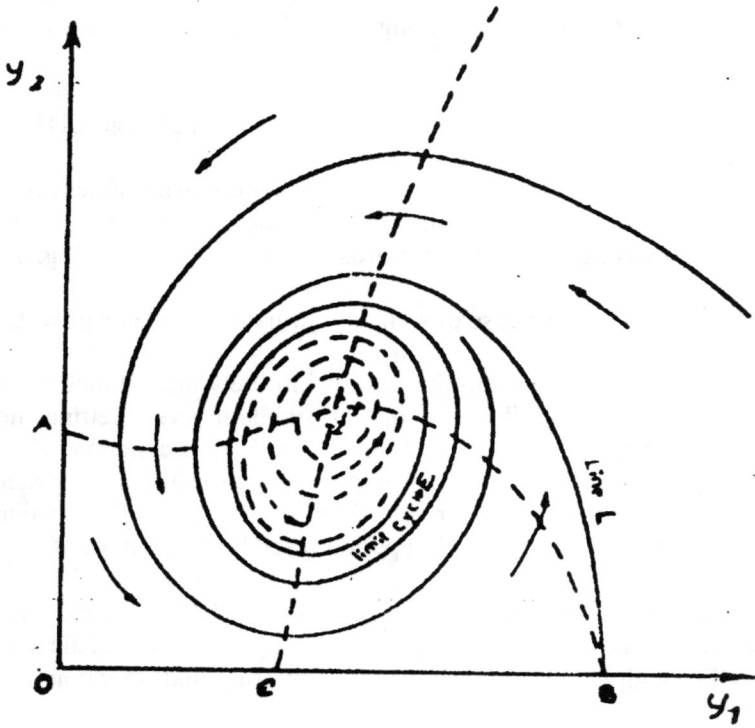

Fig. 40 Schematic phase-plane solutions around stable closed orbit E

For a more precise analysis of the behaviour of these phase-plane solutions, see Rescigno & Richardson [14] and for a discussion on a different set of predator-prey constraints see Brauer [1].

Nowadays, Kolmogorov CDS are part of a wider Poincaré-Bendixson theory of non-linear dynamics in continuous 2-D spaces. Unfortunately, we do not know how to extend it, as it is, to 3-D or higher-D spaces.

32. LANCHESTER LINEAR AND QUADRATIC CDS

We now study 2-population CDS such that each population destroys the other for supremacy until one of them is brought to extinction. Once that happens, the whole dynamic process comes to an halt and the surviving population is said to have won the "fight".

We can consider such CDS as special cases of the previous Lotka-Volterra predator-prey CDS when the destruction of the individuals of the populations is so rapid that no increase in size of either population takes place.

The linear CDS

Historically, the simplest one is a pair of differential equations

$$\begin{cases} D_t y_1 = -\lambda_1 y_2 \\ \\ D_t y_2 = -\lambda_2 y_1 \end{cases}$$

with $y_1{}^0 = y_1(t_0)$ & $y_2{}^0 = y_2(t_0)$ given and $\lambda_1, \lambda_2 > 0$ constant. Here, the "attrition" coefficients λ_1, λ_2 are called the "fighting power" coefficients since they attempt to measure the damage inflicted by the other population. To find the corresponding phase-plane solution, we re-write the above equations as

$$\begin{cases} \lambda_2 y_1 \ D_t y_1 = -\lambda_1 \lambda_2 \ y_1 y_2 \\ \\ \lambda_1 y_2 \ D_t y_2 = -\lambda_1 \lambda_2 \ y_1 y_2 \ . \end{cases}$$

Subtracting,
$$\lambda_2 y_1 \ D_t y_1 - \lambda_1 y_2 \ D_t y_2 = 0 \ .$$

Integrating, we get the phase-plane solution

$$\lambda_2 [\, y_1 \,]^2 - \lambda_1 [\, y_2 \,]^2 = K \quad \text{with} \quad K = \lambda_2 [\, y_1{}^0 \,]^2 - \lambda_1 [\, y_2{}^0 \,]^2$$

or $\qquad$ $\lambda_2 \, [\, y_1 \,]^{\,2} - \lambda_2 \, [\, y_1^{\,0}]^{\,2} = \lambda_1 \, [\, y_2 \,]^{\,2} - \lambda_1 \, [\, y_2^{\,0}]^{\,2} \; .$

Note that the weighted difference of the squares of y_1 and y_2 is constant in time and this is known as the Law of the Squares.

Since the derivatives are always negative, the values of y_1 and y_2 can only decrease from their initial values $y_1^{\,0}$ & $y_2^{\,0}$ and, at some point in time, say $t_\#$, either y_1 or y_2 becomes zero and the "fight" comes to an halt. More important, the result of the "fight" is known in advance.

In fact, if $K > 0$, y_1 wins the fight. Otherwise, y_2 is the winner. In the case $K = 0$, both lose it.

The quadratic CDS

From the previous CDS let us consider the case when the differential equations are quadratic in y_1 and y_2, i.e.

$$\left\{ \begin{array}{l} D_t\, y_1 = -\,\lambda_1 \; y_1 \, y_2 \\ \\ \\ D_t\, y_2 = -\,\lambda_2 \; y_1 \, y_2 \end{array} \right.$$

with $y_1^{\,0} = y_1(t_0)$ & $y_2^{\,0} = y_2(t_0)$ given and $\lambda_1, \lambda_2 > 0$ constant . Again, because the derivatives are always negative, the size of both populations can only decrease in time until one of them (or both) become(s) extinct. The surviving population wins the "fight".

To obtain the phase-plane solution, we re-write the above as:

$$\left\{ \begin{array}{l} \lambda_2 \, D_t\, y_1 = -\,\lambda_1 \, \lambda_2 \; y_1 \, y_2 \\ \\ \\ \lambda_1 \, D_t\, y_2 = -\,\lambda_1 \, \lambda_2 \; y_1 \, y_2 \; . \end{array} \right.$$

Subtracting,

$$\lambda_2 \, D_t\, y_1 - \lambda_1 \, D_t\, y_2 = 0 \; .$$

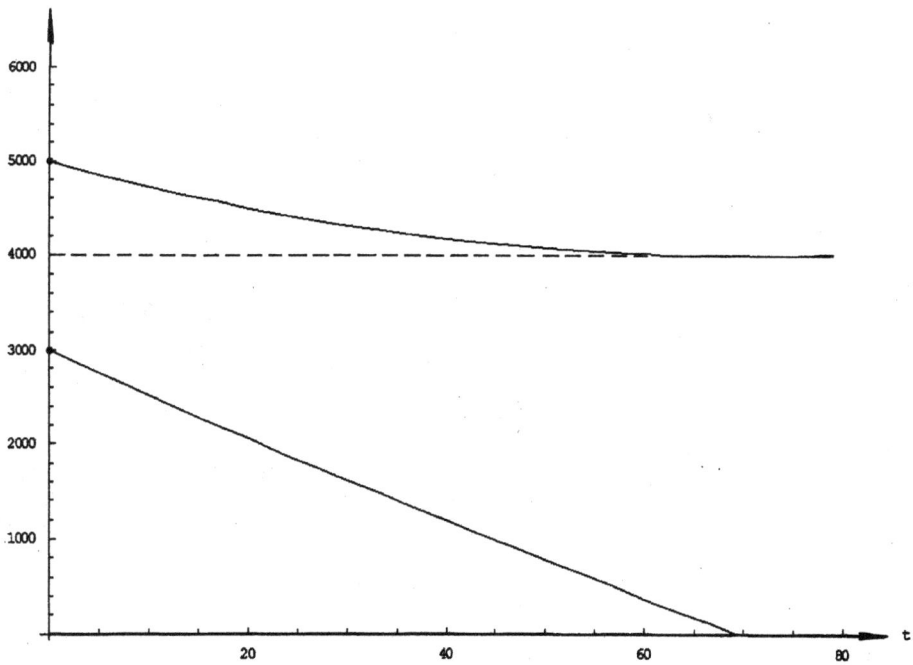

Fig. 41 Lanchester linear CDS with $\lambda_1 = \lambda_2 = 0.01$

Integrating, we obtain the phase-plane solution

$$\lambda_2 \, y_1 \, - \, \lambda_1 \, y_2 \, = \, K \qquad \text{with} \qquad K \, = \, \lambda_2 \, y_1{}^0 \, - \, \lambda_1 \, y_2{}^0$$

or

$$\lambda_2 \, [\, y_1 \, - \, y_1{}^0 \,] \, = \, \lambda_1 \, [\, y_2 \, - \, y_2{}^0 \,]$$

and the weighted difference of y_1 and y_2 is constant in time .

Again, the result of the "fight" is known beforehand and we do not need to solve the system of differential equations to discover it. Clearly, if $K > 0$, y_1 wins the "fight". Otherwise, y_2 is the winner for $K < 0$. However, to know the time $t_{\#}$ it takes for y_1 or y_2 to wipe out the adversary, we have to do it.

A comparative numerical example

Let us take: $y_1{}^0 = 5\,000$ & $y_2{}^0 = 3\,000$ as initial values for y_1 & y_2 respectively and, to make it simpler, consider $\lambda_1 = \lambda_2$.

The linear CDS

The phase-plane solution is: $[\, y_1 \,]^2 \, - \, [\, y_2 \,]^2 \, = \, 4\,000^{\,2}$.

Then, at time $t_{\#}$ $\qquad \rightarrow \qquad$ $y_1(t_{\#}) = 4\,000$ and $y_2(t_{\#}) = 0$.

Fig. 41 displays, in graphical terms, this numerical example as a function of time using standard numerical integration techniques.

The quadratic CDS

The phase-plane solution is: $y_1 - y_2 = 2\,000$.

Then, at time $t_{\#}$ $\qquad \rightarrow \qquad$ $y_1(t_{\#}) = 2\,000$ and $y_2(t_{\#}) = 0$.

Fig. 42 plots, using numerical integration techniques, this quadratic CDS as a function of time.

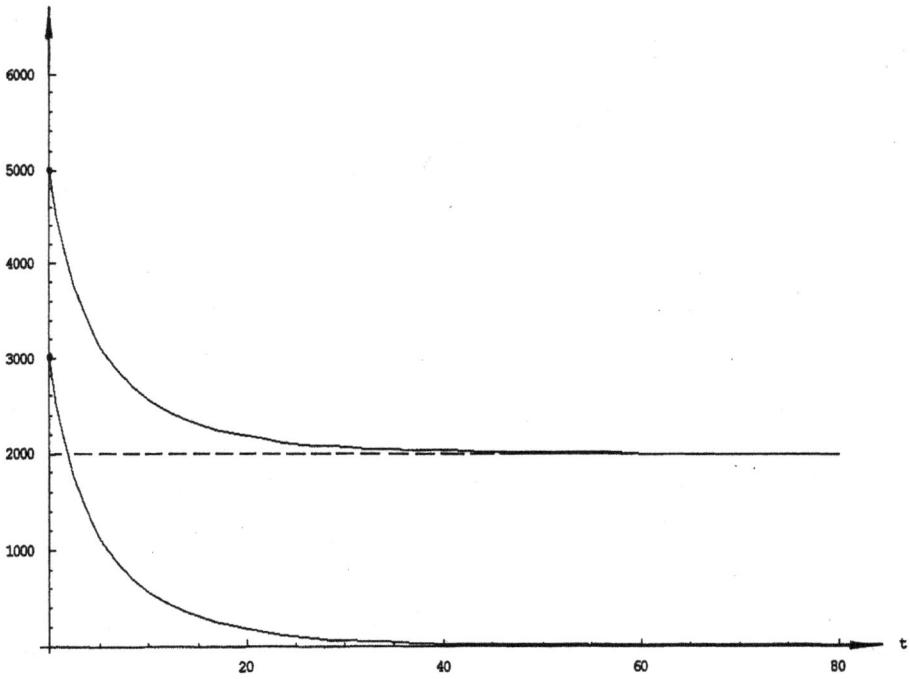

Fig. 42 Lanchester quadratic CDS with $\lambda_1 = \lambda_2 = 0.00005$

Discussion of results

In the linear CDS, the initial numerical strength of y_1 versus y_2 brings y_2 to extinction with only $5\,000 - 4\,000 = 1\,000$ y_1 casualties.

In the quadratic CDS, the situation is different: the casualties are the same for both populations (and that makes good sense since we took $\lambda_1 = \lambda_2$ in both cases) and there are $5\,000 - 2\,000 = 3\,000$ y_1 casualties.

See Lanchester [5, p39-66] for possible military meaning of these fighting CDS regarding historical battles before the advent of sophisticated weaponry and military aircraft.

More recently, a suitably modified Lanchester CDS that took into consideration troop reinforcements has been used successfully to quantify the outcome of the Iwo Jima battle in the Pacific. Reference not available.

Historically, these fighting CDS were studied before predator-prey ones were ever considered.

LIST OF ILLUSTRATIONS

BIBLIOGRAPHY AND ADDITIONAL READING

1. F. Brauer (1979) – "Boundedness of solutions of predator-prey Systems", Theor. Population Biology, 15(2), p268-273.

2. G. Fulford & altri (1997) – "Modelling with differential and difference equations", Cambridge Univ. Press, Cambridge.

3. R. A. Holmgren (1996) – "A first course in discrete dynamical systems" (2nd ed.), Springer –Verlac, N.Y.

4. A. Kolmogorov (1936) – "Sulla teoria di Volterra della lotta per l'esistenza", Giornale Istituto Italiano degli Attuari,VII, p 74-80.

5. F. W. Lanchester (1916) – "Aircraft in Warfare: The dawn of the Fourth Arm", Constable & Colin, England.

6. A. J. Lotka (1923) – "Contribution to quantitative parasitology", J. Washington Academy of Sciences, 13, p152-158.

7. A. J. Lotka (1925) – "Elements of Physical Biology", Williams & Wilkins, Baltimore. Reprinted (1956) under the title "Elements of Mathematical Biology" Dover, N.Y.

8. M. Martelli (1992) – "Discrete Dynamical Systems and Chaos", Longman UK, Harlow, England.

9. E. Martini (1921) – "Berechnungen und Beobachtungen zur Epidemiologie und Bekampfung der Malaria auf Grund von Balkanerfahrungen", Gente, Hamburg.

10. R. M. May (1973) – "Stability and Complexity in Model Ecosystems", Princeton Univ. Press, Princeton.

11. F. Oliveira-Pinto & M. Adibpour (1990) – "Analytical solutions of one-dimensional discrete dynamical systems with chaotic behaviour", Non-Linear Dynamics, 1, p121-129.

12. F. Oliveira-Pinto & B.W.Conolly (1982) – "Applicable mathematics of non-physical phenomena", Ellis Horwood with John Wiley & Sons, Chichester, England.

13. R. Pearl (1922) – "The Biology of Death" I-VII, Scientific Monthly, Lippincott, Philadelphia.

14. A. Rescigno & I. W. Richardson (1967) – "The struggle for life: I. Two species", Bull. Math. Biophysics, 29, p377-388.

15. R. Ross (1911) – "The Prevention of Malaria" (2nd ed.), Murray, London.

16. J. T. Sandefur (1990) – "Discrete dynamical systems, Theory and Applications", Clarendon Press, Oxford.

17. V. Volterra (1927) – "Variazioni e fluttuazioni del numero di individui in specie animali conviventi", Atti Reale Accad. Nazionale dei Lincei, Serie VI, Memorie della Classe di Scienze Fisiche, Matematiche e Naturali, II, p31-112.

18. V. Volterra (1931) – "Leçons sur la théorie mathèmatique de la lutte pour la vie" (Cahiers Scientifiques VII), Paris, Gauthier-Villars, VI.

ACKNOWLEDGEMENTS

It is my pleasure to thank colleagues and friends who not only helped me to produce this manuscript but also others who in a way or another motivated me into the subject or gave me the opportunity to teach it.

An old truth says that the best way to learn a subject is to teach it and that was certainly my case. So, I would like to thank:

- Prof. Brian W. Conolly, first at Saclant Research Centre in La Spezia, Italy and, afterwards, at Chelsea College of London University, UK, for introducing me to mathematical models and their applications;

- Colleagues at King's College of London University for their commitment towards my annual course in Mathematical Models and Simulation Languages that proved popular for many years amongst graduate students of the Inter-Collegiate M. Sc. in Mathematics at London University.

- Graduate students at Lusíada University, Portugal where I taught my first Discrete Dynamics course for their M. Sc. in Mathematics, who convinced me of its utility in their curriculum.

- Doctoral students at Pisa University, Italy, who with their enthusiasm motivated me to produce a written version of my lectures as an extended Discrete Dynamical Systems course.

- Special thanks to Prof. Andrea Maggiolo for his unfailing commitment to my well-being, while there and for commenting on the manuscript of the 2nd edition.

- Dr. Paulo Aguiar friend and colleague who kindly read part of the manuscript of the 1stedition pointing out inaccuracies and imperfections. Sometimes in this subject it is tricky to get it right when trying to make it accessible to a wider audience.

- Mariana who, at home, with love and dedication, made the writing of these pages a reality.

As usual, one ends saying that all errors, omissions and faults are still my own.

Cambridge CB2 1HX

2012 / 2022

INDEX

www.ingramcontent.com/pod-product-compliance
Lightning Source LLC
Chambersburg PA
CBHW071841200326
41519CB00016B/4191